SOLAR POWER CONVERSION
ENGINEERING, TECHNOLOGY, FINANCING

by
Oreste Bramanti

Foreword and legal contribution of Avv. LORENZO PAROLA

Solar power conversion: engineering, technology, financing

A Walter

Titolo | Solar Power Conversion – Engineering, Technology, Financing

Autore | Oreste Bramanti

ISBN | 978-88-91178-56-5

Youcanprint Self-Publishing

Via Roma, 73 – 73039 Tricase (LE) – Italy

www.youcanprint.it

info@youcanprint.it

Facebook: facebook.com/youcanprint.it

Twitter: twitter.com/youcanprintit

Author's profile

Oreste Bramanti: born on 30th of March 1969 in Messina (Italy), his education background was began in 1993 with University graduation in Industrial Chemistry as Technology orientation, at the Science Faculty of Messina University, where he discussed the experimental thesis based on renewable energy conversion, carried out at National Institute of Research for energy conversion and storage: "ITAE". Then, he got *Electron Microscopy Certificate* applied to the industrial catalysts, carried out in Copenaghen (DK) and, later on, a *Corporate Finance and Business Strategy Master* at Sole24Ore Business School in Milan.

With growing responsability, as technology specialist, to project leader up to manager and executive director, from 1993, he spent his professional life years in top players like BEZT, Henkel-ST, ENEL, ERG, LUKOIL & ERG Renew, GAZPROM NEFT Group - Nis Energy Block, operating in the Energy fields, mainly Power and Oil&Gas.

In those heavy industries, he was dealing with all the value chain of projects: from technical design and permitting to project engineering, financing, construction and operation.

In the specific, he was already in charge as Chief Operating Officer, Country and General Manager, International Business Development and M&A Director, Boards Member, EPC specialist, Technology and Operation Manager.

Thus, from Europe to China, from USA to Russian Federation, Mid-East and Africa, he carried out:

- all the renewable technologies and relevant conventional power projects, under international business development, M&A of several GW assets, EPCM contracting;
- problem solving of very important issues, coming from environmental impacts reduction, operations of power plants and refineries, waste/water/energy reuse management, treatment and process engineering;
- plans of energy efficiency, business strategy focused on innovation technology, revolutionary geothermal energy conversion from exhausted assets as Oil&Gas worm-out wells, green upgrading of process cycles.

Moreover, he is author of an international patent for geocoding and wrote a relevant numbers of technical works, scientific papers and books on advanced technologies, industrial processes and environmental control.

The last one was issued, as Italian series of Energy and Technology books and published by Sole24Ore, in 2012, to point out the state of art of renewable sources, concerning all the generation technologies, under title: *"Le Tecnologie delle Fonti Energetiche Rinnovabili"*, sponsored by Eon.

He is also in charge as Chairman of the international consulting and M&A Boutique "Develnaft", based in Slovak Republic, that especially assist Top Managers and Corporate Owners to arrange and implement its international business development and innovation technology.

He is a Visiting Professor at *Sole24Ore Business School* for the Master *"Environment, Energy Efficiency, Smart Cities"* and Master *"Export Management and International Project Development"*.

TABLE OF CONTENTS

✦ **Preface** by Lorenzo Parola (Paul Hastings)..8

✦ **Introduction to solar energy drivers** by Oreste Bramanti.....................10

1. **Best practice of industrial definitions**..12

2. **Solar resource**..15
   2.1) Characteristics of solar radiation
   2.2) Measurements and assessment of solar radiation
   2.3) Mathematical functions and physical laws

3. **Solar calculations for energy yield**..34

4. **Solar technologies according using**..36

5. **Solar technologies according materials**..39

6. **Photovoltaic energy: from manufacturing to operations**...................40
   6.1) Crystalline silicon (c-Si)
   6.2) Basic structure of crystalline solar cell
   6.3) Manufacturing of crystalline solar cell
   6.4) Pre-diffusion cleaning
   6.5) Etching wafer's surfaces: solid state diffusion
   6.6) Metallization
   6.7) Anti-reflective coating
   6.8) Operating principle
   6.9) Solar modules

7. **Solar plant configurations**..56
   7.1) Off-Grid systems
   7.2) Grid-Tied systems
   7.3) Components of photovoltaic plant
   7.4) Photovoltaic generator
   7.5) Electric power conversion, control system, PV inverters
   7.6) Electric storage systems

8. Thin film...70
   8.1) Thin film modules
   8.2) Thin silicon film cells and modules
   8.3) CdTe thin film
   8.4) CIGS thin film

9. CPV: concentrated PV cells...80
   9.1) CPV technology

10. Solar thermal energy...88
   10.1) Existing technologies for low and medium temperatures
   10.2) Glazed flat-plate collectors
   10.3) Structure of glazed flat-plate collector
   10.4) Covering
   10.5) Vacuum tube collectors
   10.6) Unglazed collector
   10.7) Energy source of a collector and efficiency curve

11. Solar thermal energy applications...................................103
   11.1) Sanitary hot water production
   11.2) Classification of systems
   11.3) Hydraulic circuit
   11.4) Control system
   11.5) Criteria for determining solar plant dimensions
   11.6) Technical-economic data
   11.7) Room air-conditioning
   11.8) Space solar-heating system
   11.9) Low temperature heating system
   11.10) Space solar-cooling system
   11.11) Absorption cooling groups and solar-cooling
   11.12) Storage of thermal and cooling energy
   11.13) Heat storage
   11.14) Cold storage
   11.15) Short-term heat storage
   11.16) Coexistence of radiant panels and fan-coils
   11.17) Plant control logic during summer operation

12. CSP: concentrating solar power......................................145
   12.1) Parabolic trough collector technology
   12.2) Main components

12.3) Collectors

12.4) Receiving tube

12.5) Reflective panels

12.6) Solar tracking system

12.7) Design elements

12.8) Industrial scale-up of technology

12.9) Development in advanced market: Italy's example

12.10) New applications

12.11) Economic aspects

12.12) Linear Fresnel collector

12.13) Solar tower technology

12.14) Thermal energy storage

12.15) Stirling dish technology

12.16) Solar updraft tower

13. OLED technology..190

14. Global markets and challenges of solar systems...........................192

14.1) PV, CPV and Thin film

14.2) CSP

15. PV plant and environmental impact..210

16. Polygeneration...214

17. New frontiers of R&D and key opportunities...............................218

17.1) Solar chemistry

17.2) R&D

17.3) New cells

17.4) Smart photovoltaics

17.5) PV systems and smart grid

17.6) Geocoding

18. Financing..237

18.1) Risk profiles

18.2) Best practice of FIDIC standards

18.3) Power Purchase Agreements (PPA) in the Solar Industry by Avv. Lorenzo Parola

18.4) Valuation of economics

18.5) Valuation model

BIBLIOGRAPHIC REFERENCES..258

Preface

by

Lorenzo Parola

(Senior Partner in the Corporate and Energy practices of Paul Hastings)

It has been recently reported that in 2013 only, 28.023 new photovoltaic plants (for a total capacity of 727MW) commenced commercial operations in Italy without benefitting from the feed-in tariff regime.[1] This happens simultaneously with the challenge by many investors, both before Italian Courts and Arbitral Courts under the Energy Charter Treaty, of the heinous *"spalma-incentivi"* decree under which the Italian Government cut the incentives available to PV generators retroactively.

In the solar sector reality goes much faster than regulations: while many EU Governments are trying to get away with blatant breaches of their previous commitments, players in the solar space are already playing a different game.

In solar, market-based models are increasingly surpassing incentive-based models. Alongside grid parity projects, PPAs with merchants (not only utilities but also retailers, global hotel chains, heavy industries) and residential customers (for instance, the SolarCity PPA-lease model) are becoming mainstream routes-to-market based on which banks are ready to provide long-term financing.

In this context, I have no doubt that this book, written with outstanding competence and dedication by Oreste Bramanti, will be widely welcomed both in Italy and abroad.

Indeed, compared to other books on solar energy, this work has unique features:

✓ it has an interdisciplinary approach: beside a rigorous technical and engineering analysis, a comprehensive review of both economic aspects and key bankability features of projects contracts lay the foundations for sound financial modelling of solar projects;

✓ it encyclopedically addresses the whole value chain: the analysis encompasses the features of solar radiation, material science and

---

[1] *LookOut – Rinnovabili Elettriche Q3 2014*, eLeMeNS. http://www.lmns.it/wp-content/uploads/2012/04/LooOut-RES-E-Q3-2014_table-contents.pdf

engineering, manufacturing of modules and other components, plant configuration including environmental assessment; and

✓ albeit solidly anchored in today's technological reality, it provides a visionary outlook on the new frontiers of R&D and technological applications.

My wish is that this work will become a handbook for those entrepreneurs and managers, as well as for their lenders and consultants who, building on the learning curve and best practices developed in mature markets such as Italy, Spain and Germany, will dare exploring and contaminating new markets.

Looking back we can now say that, with good approximation, coal and oil powered, respectively, the nineteenth and the twentieth century. I would not be surprised if, in one hundred years or so from now, economics and industry historians will maintain that the twenty-first century was powered by solar energy.

Milan, 30 November 2014

Avv. Lorenzo Parola

lorenzoparola@paulhastings.com

Introduction to solar energy drivers
by
Oreste Bramanti

The main factors driving so huge growth in the solar energy industry during last years include a move towards green revolution to fight GHGs (green house gases) emissions, green deals and renewable sources economy, energy efficiency programs as well as zero-energy buildings, where the end-users begin to take note of the better economics and aesthetic impacts of conventional photovoltaic systems.

Besides, to fight climate changes the global environmental issues pushed up a very important and increasing concerns about global warming, sustainability and consumer drive to contribute to a cleaner environment and reduce reliance on fossil fuels.

Furthermore, rising electricity costs and consumer sensitivity drive to offset electricity bills by providing some or all of their own electricity for on-site consumption and/or to feedback to the grid.

In addition, the basic legislative requirements and targets for reducing of carbon emissions have shown a significant impact on the international legal and regulatory frameworks.

In that way, public opinion and industrial interests found out a good fitting phase, focused on relevant incentives and specific support through implementing new energy authority regulation and special tariff systems to promote that solar power generation.

Currently, subjects of current Climate Change Challenge are focused on:

✓ create more energy generation, securing its increasing demand and lower CO_2 emissions.

✓ As party to the UNF Convention on Climate Change and old Kyoto Protocol, the EU-27 committed to reduce emissions by 7.8% compared to 1990 levels.

✓ Market opportunities and development become obligations under EU according to Dir. 2009/28.

✓ Governments across the world are making a strong push to make RES a significant source of electrical energy.

✓ Nuclear decline/phase-out will push on alternative-sources demand growing.

✓ U.S. president Barack Obama's plan is to have 25% of US electricity coming from renewable energy sources by 2025.

- ✓ Chinese government is raising the country's wind power capacity to 100 GW by 2020 representing annual growth rates of 20%.
- ✓ There are real perspectives in international markets of renewable power generation, due to a wide availability of resources (wind, geothermal, hydro, solar, biomass).
- ✓ Technology innovations will meet requirements of new entrants (ex: grid parity).
- ✓ Next years will be a great window of business opportunity on global basis, in which legal and regulatory frameworks will have to support innovation of technologies, improve electrical grid connection, upgrade solar cells, inverters, battery banks and conversion processes, push up grid parity to spread additional industrial players, independent power producers and developers through specific market conditions, new incentive and financing schemes.
- ✓ In the past, it was clear that all the PV value chain was based on the leading key drivers for marketing:

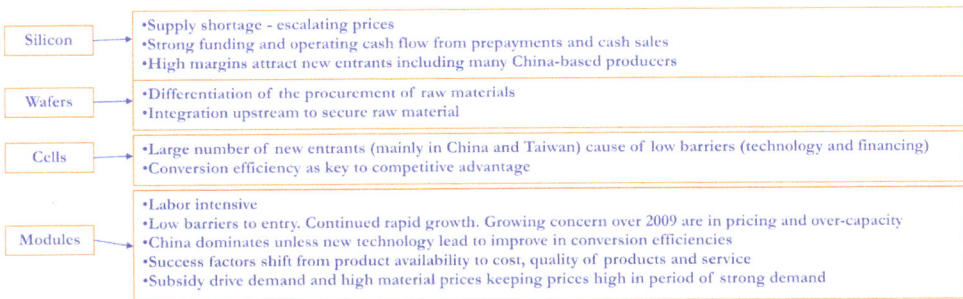

Silicon	•Supply shortage - escalating prices •Strong funding and operating cash flow from prepayments and cash sales •High margins attract new entrants including many China-based producers
Wafers	•Differentiation of the procurement of raw materials •Integration upstream to secure raw material
Cells	•Large number of new entrants (mainly in China and Taiwan) cause of low barriers (technology and financing) •Conversion efficiency as key to competitive advantage
Modules	•Labor intensive •Low barriers to entry. Continued rapid growth. Growing concern over 2009 are in pricing and over-capacity •China dominates unless new technology lead to improve in conversion efficiencies •Success factors shift from product availability to cost, quality of products and service •Subsidy drive demand and high material prices keeping prices high in period of strong demand

As we will see in the next pages, solar modules can be arranged into arrays, that are large enough to function as successful power stations converting sunlight into electrical energy for industrial operations, commercial buildings, military factories and residential end-users.

Thus, solar modules in smaller configuration and capacity should be installed on buildings for residential or commercial use as well.

Solar panels can also be operated in remote areas where there is a short supply of electricity or where electricity cannot be delivered such as in space.

Finally, solar conversion technologies shall be entrusted to drive the next power revolution, based on energy storages and smart grids or districts.

1. Best practice of industrial definitions

Basically, there is one and only best practice of definitions to be managed by operating and management team for both as the final scopes of work and during all the phases as project engineering, project development and project financing.

Any professional that followed up and made a solar deal, until the project financing closing, could report what technical definitions and bankability terms really need for the project building on success industrial basis.

Beam Radiation: The solar radiation received from the sun without having been scattered by the atmosphere. (Beam radiation is often referred to as direct solar irradiance, DNI).

Diffuse Radiation: The solar radiation received from the sun after its direction has been changed by scattering by the atmosphere. (Diffuse radiation is referred to in some meteorological literature as sky radiation or solar sky radiation).

Total Solar Radiation: The sum of beam and the diffuse solar radiation on a surface.

Irradiance (W/m²): The rate at which radiant energy is incident on a surface per unit area of surface. The symbol G is used for solar irradiance, with appropriate subscripts for beam, diffuse, or spectral radiation.

Irradiation (J/m²): The incident energy per unit area on a surface, sound by integration of irradiance over a specified time, usually an hour or a day. Insolation is a term applying specifically to solar energy irradiation. Symbols **H** and **I** are used for insulation for a day and an hour, respectively.

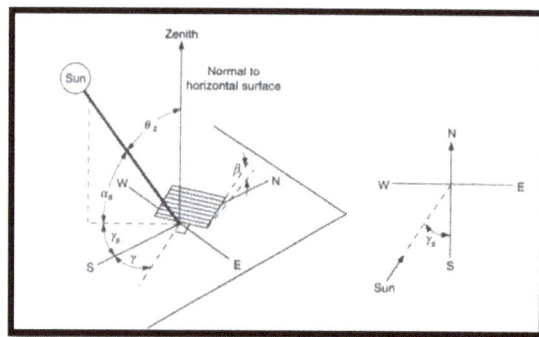

Picture 1.a - Angles and axes of sun radiation

ϕ - **Latitude,** the angular location north or south of the equator, north positive; $-90° \leq \phi \leq 90°$

δ - **Declination,** the angular position of the sun at solar noon (i.e., when the sun is on the local meridian) with respect to the plane of the equator, north positive; $-23.45° \leq \delta \leq 23.45°$.

β - **Slope,** the angle between the plane of the surface in question and the horizontal; $0° \leq \beta \leq 180°$. ($\beta > 90°$ means that the surface has a downward-facing component).

γ - **Surface azimuth angle,** the deviation of the projection on a horizontal plane of the normal to the surface from the local meridian, with zero due south, east negative, and west positive; $-180° \leq \gamma \leq 180°$.

ω - **Hour angle,** the angular displacement of the sun east or west of the local meridian due to rotation of the earth on its axis at 15° per hour; morning negative, afternoon positive.

θ - **Angle of incidence,** the angle between the beam radiation on a surface and the normal to that surface.

Additional angles are defined to describe the position of the sun in the sky:

θz - **Zenith angle,** the angle between the vertical and the line to the sun, that is, the angle of incidence of beam radiation on a horizontal surface.

αs - **Solar altitude angle,** the angle between the horizontal and the line to the sun, that is, the complement of the zenith angle.

γs - **Solar azimuth angle,** the angular displacement from south of the projection of beam radiation on the horizontal plane, shown in Picture 1.a. Displacements east of south are negative and west of south are positive.

Some solar collectors "track" the sun by moving in prescribed ways to minimize the angle of incidence of beam radiation on their surfaces and thus maximize the incident beam radiation.

The angles of incidence and the surface azimuth angles are needed for these collectors.

Tracking systems are classified by their motions. Rotation can be about a single axis (which could have any orientation but which in practice is usually horizontal east-west, horizontal north-south, vertical, or parallel to the earth's axis) or it can be about two axis. The performance of a single axis concentrating solar collector depends on the tracking axis.

Picture 1.b - Single axes tracking system

2. Solar Resource

2.1 Characteristics of solar radiation

Solar radiation is a fundamental source of the Earth's "energy system"; in fact, that's the other source being the heat obtained by radioactive decay in the planet's core and, thus, it is the basis of all natural cycles and manifestations of life, including many human activities.

Understanding how it interacts with the atmospheric layer surrounding the Earth, before it reaches its surface, is a prerequisite for understanding natural phenomena regarding climate and meteorology, as well as for locating and designing systems using solar energy.

In space, the Sun emits energy mostly in the form of electromagnetic radiation. This emission is constant in time, equal in all directions (isotropic) and amounts to tremendously great quantities: in terms of power, it has 3.85×10^{26} W (385 billion billion megawatts), eradiated in the entire solid angle of 4 sr (steradians).

Solar radiation is the result of nuclear reactions in the Sun's core. In fact, in the moment of its formation, a star is a gas mass, predominantly consisting of hydrogen, the lightest element in nature; at the first stage of a star's life, hydrogen atoms are *fused* into helium and subsequently helium is transformed into other, each time heavier elements.

All of the mentioned reactions are exothermic and, thus, generated energy is transferred outside the core, to a convective layer of the star, called *photosphere*, wherefrom it is irradiated into interplanetary space.

The Sun is a young star, since it was formed slightly more than four billion years ago, and it is still at the stage of converting hydrogen into helium.

This conversion will be stable for another several billion years: hence, comparing to the time scale of the human civilization, the Sun constitutes a source of energy that may be considered "eternal".

The celestial bodies of the Sun system, especially planets including the Earth, receive a fraction of the energy irradiated by the Sun into the interplanetary space in a measure proportional to the solid angle under which each one is seen from the Sun; the angle depends on the planet size and its distance from the Sun. In case of the Earth, the 'section' available for receiving solar radiation approximately corresponds to a circle of an average radius of about 6367 km. Its distance from the Sun, varies throughout the year since it has an elliptic orbit; at an average distance, slightly less than 150 million km (^{2}), a plane perpendicularly (normally) exposed to the solar radiation, on top of the atmosphere, will be exposed to the following radiation per unit area:

$$i_{sc} = 1366.9 \ \text{W/m}^2$$

that is called *solar constant*^{3}. Technically, the power of radiation incident on the unit area is called irradiance, and its unit is W/m^2 (or its multiple). On the soil, irradiance different from the solar constant is measured and it varies any moment due to numerous factors:

a) the distance Earth-Sun (*);
b) the orientation of the incidental plane with respect to solar radiation;
c) the attenuation caused by the atmosphere molecules.

Variability of the distance Earth-Sun is taken into account to correct the solar constant using the so-called *factor of the Earth's orbital eccentricity*.

It varies a few percents per year: its maximal value of 1.035 is reached on January 3rd, when the Earth passes through the point closest to the Sun (perihelion), while its minimal value of 0.967 occurs between July 3rd and 4th, at the maximal distance from our star (aphelion).

Irradiance depends also on the orientation of the receiving surface. For equal powers per unit solid angle (in W/sr) transported by radiation, if the receiving surface is normal to the direction of solar radiation, the surface exposed will be the smallest and the irradiance (in W/m^2) will be maximal.

Generally, solar radiation will form an angle ϑ, different from zero, to the normal, called the *angle of incidence* (Fig. 2.a): in that case, irradiance will be attenuated by a factor that amounts cos ϑ.

The solar constant, corrected by the factor of eccentricity and the cosine of the angle of incidence is called *top of atmosphere solar irradiance*. The orientation of the receiving surface is arbitrary, but there are two positions of particular interest: *normal*, already

---

*) *More precisely: 149597890 km.*

To get an idea how great it is, consider that a surface extending to about 80 km^2 (i.e. a square with a side of 7.7 km, less than a five-thousandth part of the entire territory of Italy) receives the total electric power installed in Italy in 2011, which amounts approximately 110 GW (GW = gigawatt = a billion watts, of course).

mentioned, for which the irradiance is maximal, all other conditions being equal, and *horizontal*. In this case, the receiving plane is fixed parallel to the Earth surface.

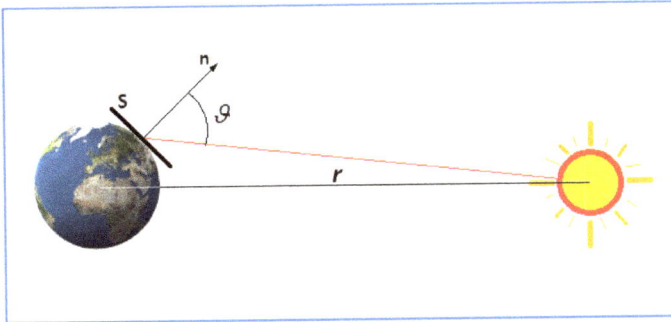

Figure 2.a – Earth-Sun distance and the angle of incidence of solar radiation

The angle of incidence of solar radiation with respect to a horizontal surface is called *zenith angle* and is usually marked by ϑ_z. Since it depends on the position of the Sun, the term $\cos \vartheta_z$ expresses its arc movement during the day: its value increases during the morning to reach maximum at noon, and decreases after noon (in the dawn and at the sunset its value is zero).

The same trend will be noticed with the top of atmosphere horizontal irradiance, which it the function of $\cos \vartheta_z$ (Fig. 2.b).

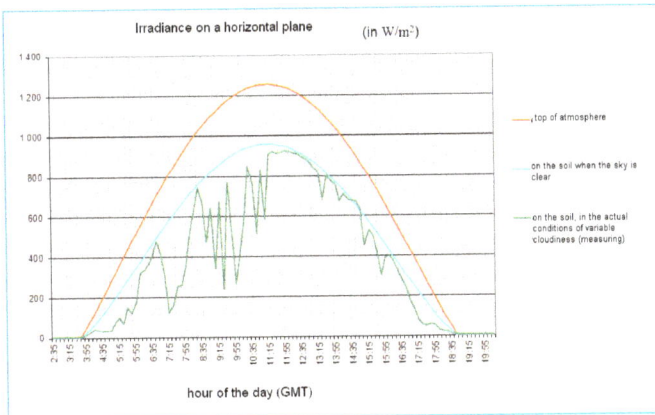

Figure 2.b – Daily profile of irradiance on a horizontal plane

All the quantities examined so far are irradiance, i.e. the *power* of radiation per unit area. In case one looks for the radiation energy per unit area in a finite time interval, irradiance needs to be expressed with respect to time.

The radiation energy of the solar origin, incident on a unit area, is called solar irradiance.

In the SI, the solar irradiance unit is J/m^2 (Joule per square meter), but units such as MJ/m^2 (megajoule per square meter) and kW/m^2 (kilowatt hour per square meter) are also in use.

In fact, the unit conversion is as follows: $1\ kWh/m^2 = 3.6\ MJ/m^2$.

Although the definition of irradiance applies to any interval, two intervals are currently in use: *hourly and daily*.

The calculation of the top of atmosphere irradiance, and consequently irradiance on the soil – is influenced by various factors: the position of the Sun in the sky, which changes during the day and in the course of the year, geographic position, and finally, the orientation of the receiving surface.

The effects relating to these factors cannot be easily explained. Normal irradiance is greater in winter in all the parts of the globe, since in that period the Earth is closer to the Sun than in summer.

On the other hand, the Earth's rotation axis is such that in the latitudes of the northern hemisphere, where Europe is located, the Sun rises in the sky at noon considerably higher than in other periods and hence has a longer path (the day lasts longer).

This latter effect is greater than the former one, which acts in the opposite direction, so that in summer, even though the normal top of atmosphere irradiance is smaller, the horizontal irradiance is considerably higher in all intervals of the day, as well as the integrated daily irradiance value.

The figure 2.c indicates the annual tendencies of top of atmosphere irradiance for certain orientations of receiving surfaces.

The figure represents a specific location, but the trend is qualitatively similar in any geographic position.

For stationary surfaces, the best performances can be obtained when they are oriented south and inclined with respect to the horizontal plane at an angle comparable to the latitude of the site.

In power plants using the solar source, the devices collecting the energy transported by solar radiation are installed on the soil level, both in case of photovoltaic and thermal modules, and in case of reflecting mirrors in a concentrated solar power facility.

In all these cases, the irradiance available on the soil is smaller than the one on top of the atmosphere, because it is affected by factors that, combined, lead to attenuation of the net effect (Fig. 2.c):

a) while passing through the atmosphere - whose thickness is approximately 80 km perpendicularly; which is so thinned on top that in our latitudes it is irrelevant for purposes of power - a part of solar radiation is absorbed or scattered due to the collision with the atmospheric molecules (including water vapor, aerosols and clouds).

   Attenuation affects all the wave lengths, but not to the same extent, since some of the molecules (oxygen, carbon-nitride) absorb light selectively.

   Direct solar radiation i.e., one that consists of solar radiation that passes through the atmosphere layer undisturbed, reaches the soil from a single direction, that changes constantly during the day, since it is defined by the position of the Sun in the sky;

b) if not absorbed, a fraction of solar radiation that collides with the atmosphere is scattered in all directions: a part of it goes back to space, and a part of it reaches the soil.

   The latter is called *diffused solar radiation* and, obviously, it comes from all directions of the celestial hemisphere.

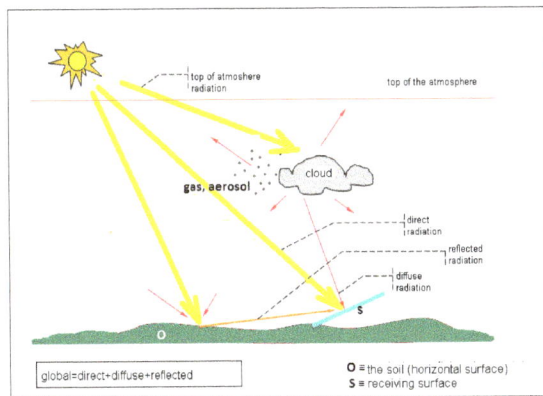

Figure 2.c – Decomposition of radiation at the soil to its components

In a given moment, *global* solar irradiance on a horizontal surface is defined by the sum of the direct and diffuse components, while a surface inclined with respect to the horizontal plane, apart from these two components, also includes the *solar irradiance reflected* by the soil.

In some situations, the exposure of a site varies to a considerable extent due to many factors including the day and night alternation, seasonal cycles and changes in meteorological conditions.

At our latitudes, in most favorable conditions, the attenuation of beam irradiance due to their passing through the atmosphere can be 15-20% and it is possible to register direct normal irradiance values ranging 1000-1500 W/m² on the soil.

Direct normal irradiance is often shortened as *DNI*.

Models of Radiation Transmission in the Atmosphere represents a very important section that offers a schematic overview of various factors that contribute to attenuation of direct irradiance in the atmosphere:

1) effective length of the solar radiation course;
2) type and density of the gas molecules in the air encountered on the course;
3) quantity of humidity (water steam) and aerosols (solid pollution particles);
4) cloud intensity.

In the technical literature, there are many empirical models relating the irradiance at the soil and the above mentioned factors.

The examined irradiance is usually DNI, the one diffused on a horizontal surface and the global one (always on a horizontal surface), or daily, hourly and daily and monthly average.

The sky without clouds is referred to as *clear*. In all of the proposed models of the clear sky, the first two factors from the above list are characterized by two parameters: *relative air mass* and *Raleigh factor of fading*.

However, even when the sky is clear, the atmosphere contains various quantities of humidity and aerosols (number 3).

All other conditions being equal, 'clear' atmosphere, which attenuates solar radiation to a lesser extent, is the atmosphere without humidity and aerosols *(dry and clean air)*.

However, in order to express the attenuation that reflects the actual condition of the atmosphere, *Linke turbidity factor* is used, which is assumed to be equal to 1 in case of dry and clean air, proportionally increasing with the load of humidity and aerosols. Anyway, if the mentioned anomalies are left aside, this value varies slowly over time and little in space; therefore, the useful tables of the average monthly values have been prepared for many places in Italy and all over the world.

The presence of clouds (number 4) is the most accidental and the most intensive phenomenon in the atmosphere.

For purposes of comparison, Figure 2.c shows the daily profile of irradiance on a horizontal plane for three different conditions:

1) on top of the atmosphere,
2) when the sky is clear, and finally,
3) the actual situation on a June day in Rome, with variable cloud cover.

The trend visible in the third case differs from the previous one with respect to the phenomenon of the cloudiness in the atmosphere appearing during the day; for several moments, when the sky was clear, the two profiles had the tendency to coincide.

All the time, solar radiation on an *Inclined Plane* must be valuated to select the most efficient plant solution and configuration.

The simplest solution for a facility using a solar source of energy consists of horizontal exposure of the receiving surface.

However, it has been shown (Fig. 2.c) that such orientation is not the best one for purposes of collecting the energy transmitted by the incident solar radiation, hence the effectively adopted design solutions provide receiving surfaces inclined in an adequate way.

The chosen inclination takes into consideration the geographic position, operating period, and, naturally, the design and micro-siting restrictions.

The best configuration corresponds to the plane normal to the radiation, but this solution imposes demanding construction options: the surface will not be stationary, but has to follow the Sun position every moment; nor can the rotation axis of the tracker be stationary, so the design solution has to provide the possibility of rotating around two axes.

The best compromise adopted in practice for medium- and large-size facilities, both for the photovoltaic and solar thermodynamic facilities, consists of the receiving parts rotating around a fixed axis, usually in a horizontal position along the direction north-south.

Alternatively, especially in case of small facilities, a stationary surface is installed, choosing the direction that would be the best in most operating circumstances.

Usually, the best orientation is toward south and inclined at angles that do not differ much from the latitude of the site.

Obviously, this suggestion cannot be followed in all cases: a case in point is a façade of a building, vertical in simplest cases.

In summer, such surfaces are not exposed due to the fact that the considerable elevation of the Sun on the Earth's horizon translates into low elevation on the façade horizon, for the most part of the day (in relation to this, refer to the graph in Figure 5.3).

According to the norm, irradiance measurement data on the inclined surfaces are not available: it is impossible to conduct continuous measurements for all orientations and all sites of interest.

This inconvenience can be eliminated if the data on global irradiance on horizontal plane are used – the banks of the meteorological data are usually supplied with them – in order to apply specific calculations on inclined surfaces, pre-selected from time to time.

This procedure requires the knowledge of all modalities of global radiation fractioning to its direct and diffuse components, diffuse properties of the sky and reflexive properties of the soil.

To that end, adequate empirical models are required.

According the terminology of Radiometric Quantities I have to highlight that physical quantities describing complex phenomena related to the emission, propagation and reception, with consequent absorption, reflection and/or transmission are called radiometric quantities.

Since radiometric quantities have been used in various areas of science and technology, various terms have come to denote the same quantity.

The case in point is the term *radiant exposure* which, in the solar context is referred to as irradiance and, more often and generically, as radiation.

In fact, *solar radiation* is a commonly used expression, denoting either the physical entity of radiation which consists of electromagnetic waves emitted by the Sun and interacting in various manners with the atmosphere and soil, or, in a more specific meaning, denoting a physical quantity, synonymous with irradiance: it is common to consider solar radiation per hour, daily, monthly, annually etc.

However, the same expression is used to denote other physical quantities, related to the radiation phenomena, such as irradiation, which is sometimes called *instantaneous* solar radiation.

In order to determine which quantity it refers to, it helps to use the unit of measurement (when there is one and when it is correctly indicated):

- MJ/m^2 or kWh/m^2 is used for irradiance (as radiation energy per unit area in a certain interval).
- W/m^2 is used, on the other hand, for irradiance equal to radiation power per unit area.

2.2 Measurements and assessment of solar radiation

Quantities to be measured and the necessary instruments are simply the proper valuation tools of the right solar system.

Solar radiations can actually denote three quantities:
- ✓ global radiation on a horizontal plane,
- ✓ diffused radiation on a horizontal plane,
- ✓ direct normal irradiance.

The insufficient availability of the data on direct irradiance should also be mentioned (with intervals that should be at least hourly), both in terms of number of sites and in terms of periods of data acquisition, which is not the case when measuring the global radiation on a horizontal plane.

The main project of a solar facility, especially for a concentrated one, cannot be considered without the knowledge of these parameters.

Apart from being essential for calculations of energy that a given facility can produce, hourly trends allow identification of possible critical points in the system for control and regulation and thermo-mechanical stress, to which parts of the system are exposed.

Finally, the knowledge of the direct irradiance profile allows determining correctly the size of the accumulation system, based on technological and economical considerations.

To overcome this gap, an advanced Country has installed several measuring stations on the its territory, building an *Actinometric network*.

To give an example of that best practice, Figure 2.d shows location of five stations on the Italian territory, currently operating, while Table 2.e shows the data for each of them – station activation date and measurements to be performed.

The stations are distributed in places envisaged for building solar thermodynamic facilities (Specchia, Montalto di Castro, Priolo Gargallo) or for example in the ENEA experimental centers (Casaccia, Trisaia).

The data acquired through the network are stored in a data base that can be used for syntheses and statistics (hourly, daily data, daily average, monthly etc.); apart from that, the data are used for determining the parameters that play a role in models of radiation estimation, starting from the satellite data.

The most critical element for measuring direct solar irradiance is the solar tracker, a moving part that performs a double task: to shield the diffuse radiation sensor on the horizontal plane using an adequately oriented black metal sphere and to direct the pyrheliometer, an instrument for measuring direct irradiance using a solar disc. Thus, the tracker needs to be very precise and reliable in order to minimize the unadjusted periods, when the quantities measured are false.

Figure 2.d – The ENEA Actinometric Network: map of locations with installed stations for data acquisition

As follows, one may found out a significant case of Actinometric network stations (Table 2.e), according the italian best practice, made by ENEA:

Name	Coordinates		Operation start date	Obtained parameters
	Latitude (North)	Longitude (East of Greenwich)		
Casaccia (RM)	42°03'	12°18'	November-01	radiation, temperature, wind
Trisaia (MT)	40°10'	16°39'	February-02	radiation
Montalto di Castro (VT)	42°22'	11°31'	February - 02	radiation, temperature, wind, humidity
Specchia (LE)	39°57'	18°16'	May-02	radiation, temperature, wind
Priolo Gargallo (SR)	37°08'	15°13'	July-03	radiation

In the further Figure 2.f, it's significant to better show a specific particular of such as actinometric station's solar tracker:

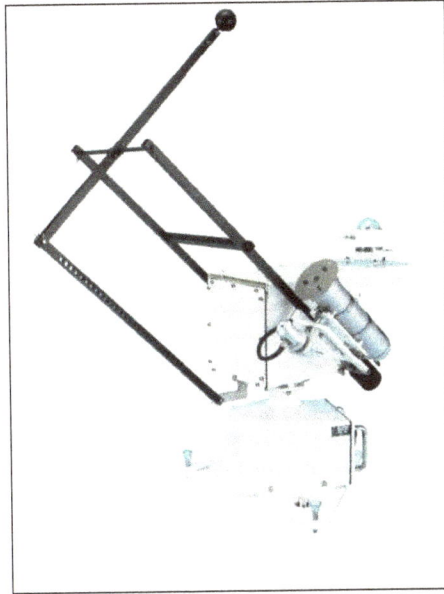

Figure 2.f - Actinometric station's solar tracker

The system orients automatically on the solar disc during its daily and seasonal course. Since the tracking has to be performed both in clear and cloudy weather, solar tracker is programmed to orient according to the solar position, which is calculated based on the hour, day and the geographic position of site (latitude and longitude).

Pyrheliometer, installed on solar tracker, consists of a thermopile, whose hot junctions are exposed to direct irradiance, while cold junctions are protected from radiation. The voltage produced by the thermopile is proportional to direct radiation.

In order to consider only the direct radiation from solar disc, the sensitive element is protected with a metal cylinder, with an internal absorbing coating.

At the end of the cylinder, directed to the Sun, there is a quartz window that insulates the system from the outside.

To measure diffuse radiation pyranometer is used, whose operating principle is identical to the one in pyrheliometer.

In this case, however, since it is required to collect radiation coming from a hemisphere, the exposed part of the thermopile is insulated with two quartz hemispheres.

In order to measure diffuse components of radiation, the sensor is required to be protected from the beams coming directly from the solar disc.

This is achieved by inserting a small sphere between active parts of pyranometer and the Sun. Series of levers and a metal sphere of 10 cm in diameter mechanically connected to the tracker, fulfill this role.

Finally, global radiation on a horizontal plane is measured by another pyranometer, completely identical to the previous one, parallel to a horizontal plane without obstructions, that can receive all the radiation regardless of its direction.

The station is equipped with an automatic system for data acquisition, which memorizes the average value and standard deviations of each quantity starting from the data measured in the previous 5 minutes.

All the data are saved on an internal memory module, which can save the data collected in an interval of several months. In regular intervals (usually weekly), the data are transmitted, and after checking if they match, they are permanently stored in the data base for further processing.

The transmission is performed using the transmission system over GSM (Global System for Mobile communication).

Besides, the main issues relating to measuring of direct solar radiation are to be underlined.

All of the installed stations permit measuring three basic quantities, for purposes of actinometric characterization:

➢ *global* radiation on a horizontal plane I,
➢ *diffuse* radiation on a horizontal plane I_d
➢ direct solar radiation on a horizontal plane I_{bn}.

The three types of measured data are related as follows:

$$I_{bn} \cdot \cos \vartheta_z = I - I_d$$

where ϑ_z is an angle formed between solar beams and the normal to a horizontal plane (*zenith angle*), which can be calculated over the site's latitude and longitude and the 'actual' hour and the day.

This method enables a quantity to be measured even in case there is an error with it, using the correct measuring of the other two quantities.

Measuring of the direct solar radiation can thus be performed with two methods: the first one using pyreliometer and the other one using two pyranometers.

In the latter case, after measuring global solar radiation on a horizontal plane as well as diffuse solar radiation using two pyranometers, the difference between these two radiations determines the direct solar radiation calculated on a horizontal plane.

Tens of thousands of data, acquired from the measuring stations, are processed each hour and, subsequently, used to perform the comparison between the direct solar radiation on a normal plane measured by a pyrheliometer with the direct radiation, obtained through calculating the difference between measurements of the two piranometers.

Finally, corrective coefficients are determined that will be applied on global radiation and diffuse radiation on a horizontal plane for the cosine of the zenith angle, in order to obtain the direct radiation measured by pyrheliometer.

This angle determines the height of the Sun in the sky and indicates time of the day, depending on geographic coordinates, the day of the year, ranging between 0° and 90°, the maximal value being 90° when the Sun is on the horizon and the minimal at noon.

Direct measuring with a pyrheliometer has three disadvantages: it requires frequent calibration, it receives radiation over glass small in size (usually 1.9 cm in diameter) and it is an expensive instrument.

The need for frequent calibration does not allow the uncontrollably long use of the instrument (preferably for months or even years), while the small dimensions of the glass and its plane surface are exposed to various inconveniences (e.g. birds' excrements) that may reduce its transparence and require operator's intervention.

On pyranometers, on the other hand, the dome-shaped cover removes foreign bodies, deposited there accidentally, through rain.

The pyranometer's stable measuring allows its autonomous use over a longer period of time than pyrheliometer, and, finally, its cost is much lower than pyrheliometer's.

If one is equipped with two pyrheliometers, one of which includes solar tracker for moving the sphere, they can determine direct radiation over the difference between measurements of the two instruments.

However, this method induces a systemic error (which is not underlined by the pyrheliometer makers) due to the fact that measuring the diffuse radiation on a horizontal plane does not take into account the fraction of diffuse radiation coming from the circumsolar area, received by the sphere that shadows the Sun.

This is demonstrated by the fact that in all stations radiation measured by the pyrheliometer shows lower figures than calculated direct radiation.

This is most probably because the diffuse radiation received by the sphere that shadows the Sun has not been considered.

The diffuse radiation consists of three components. The first one is the isotropic radiation (but a fraction of this radiation, received by a rotating sphere, is too low to produce any appreciable effect).

The other component is the radiation that comes from the horizon (the same consideration applies to this radiation as to the isotropic radiation).

The final component is the circumsolar diffuse radiation.

This is the only component that may affect the measuring and it is, to a large extent received by the sphere that shadows also the radiation coming directly from the Sun.

Thus, solar radiation forecasting results the focal point of a reliable and really suitable business valuation for investors and project financing as well: at the end, all the bankability process and approval strongly depend on the real financial modellization, meaning a proper cash-flows calculation, on realistic basis, coming from detailed solar resource assessment.

In the past several years, Italy or Spain as well have seen the expansion of solar system installations, especially photovoltaic systems, to such an extent that today Italy is in the top-class of countries in the world, after Germany, regarding the number of photovoltaic systems installed.

In fact, there are almost 350.000 photovoltaic systems of the total capacity of approximately 12 GW.

Of them, 30.000 are systems over 50 kW with the total capacity of over 9 GW, corresponding to 75% of the installed photovoltaic system capacity on the Italian territory.

In particular, the sharpest increase is noticed in large systems, i.e. over 1 MW. Therefore, there are two requirements: one at a local level, where the owners of individual systems benefit because they know in advance how much radiation they can 'collect' so they can use their facility in an adequate way, e.g. activate the program for maintenance on days when less solar energy is forecasted, or send the power produced to the grid in periods of higher compensation, bearing in mind the tariff system of the energy market; the other requirement is on a national level, where the energy Administrator has to manage all the users, implying that he/she needs to anticipate grid demands in various parts of the country.

This scenario shows that there is an evident benefit from developing solar radiation forecasting, which is a natural evolution of meteorological forecasting.

The evolution that has proven itself in the area of meteorological forecasting has allowed achieving a great extent of reliability.

It owes to an immense development of computers and the Internet, making the results of the forecasting models available to a large public, from professionals to ordinary end users.

Talking about the history of meteorological forecasting is beyond the purpose of this text; it is necessary to remember only that meteorological forecasting requires an international network for collecting and centralization of the data measured on the soil and on various altitudes all over the Earth, using powerful computes that process this data, providing solutions for complicated mathematical equations (differential equations and partial derivatives) that simulate atmospheric behavior.

Input data include the temperature values, humidity, wind and air pressure.

The data processing gives derived quantities as a result, such as cloudiness, precipitation etc. On the whole earth, there are 40.000 stations in total, including those performing measurements on the soil and in higher altitudes, coordinated by the World Meteorological Organization.

On an international level, there are several computer centers for meteorology. As for Europe, there is, for example, ECMWF (European Centre for Medium Range Weather Forecast), where a model has been developed representing the atmosphere divided in 60 vertical levels, up to 70 km in height, and 40 km in horizontal resolution. Apart from models on a global level (GM=Global Model), there have been developed models at a local level (LAM=Limited Area Model) taking into account the detailed characteristics of the site it applies to, e.g. vegetation and orography.

Models on a global scale are used to initiate those on a local level, whereby the latter ones use the output data of global model as the initial data at the entry and as the data on the borders of a specific area under consideration.

The evolution of the meteorological forecasting has resulted in developing evermore specialized models, including those forecasting solar radiation.

These models may be conceptually represented in two successive stages.

The first stage is the simulation of atmospheric movement (thermodynamics) and simulation of water transformation (clouds, precipitations...), obtainable by a series of prognostic models, similar to models used for weather forecasting.

The second one consists of the radiative transfer model, which, using the atmospheric variables on the vertical, forecasted by the first model, calculates various components of radiation on the soil for each point.

Solar radiation forecasting, developed by ENEA, was developed from the need to know in advance how much solar power will be available in a short period (today, tomorrow and the day after tomorrow) in order to guarantee an adequate administration of the solar facilities for generating electricity.

Solar energy is characterized by an intrinsic unpredictability, and, except for the interruptions due to alternation of day and night and seasonal changes, causing the daily variations of the solar power availability throughout the year, it is also affected by accidental changes in meteorological conditions.

The situation may occur when the facility is affected by an occasional passage of clouds, which cover the site of the facility partially or completely in days which are otherwise sunny, or there may occur situations when bad weather lasts for many days, when the system does not receive enough solar energy.

To program the electric power generation and to manage the energy accumulated in solar systems, it is of greatest importance to estimate the facility's productivity in the following days and hence, to provide solar radiation forecast.

The benefits from solar radiation forecasting should not be underestimated, because it could enable saving the costs of buying the accumulation equipment, which is very expensive. Actually, forecasting, properly included in facility administration programs, could reduce to a minimum the necessary accumulation.

To perform forecasting, the italian system uses suitably selected Internet sites, specialized for meteorological forecasting.

They have to be reliable, both regarding the regularity profile of their functioning and regarding the credibility profile of the meteorological forecast they provide.

Currently, the system developed by ENEA, on a daily basis and on a demonstrative level, performs forecasting for global and diffuse solar radiation on a horizontal plane, and the direct one on the plane normal to the solar beams, with a temporal resolution of an hour, for five locations in Italy: Milan, Casaccia (Rome), Monte Aquilone (Manfredonia, Foggia), Portici (Naples) and Priolo Gargallo (Siracusa).

Radiation on inclined planes can anyway be deduced from radiations derived employing adequate formulae.

The supplied radiation has to fulfill functional requirements of both solar photovoltaic and thermal systems, using global radiation and concentrated systems, using the direct radiation.

Since the knowledge of climatology of the site for which the forecasting will be performed is on top of the ENEA system and it is included in the cloud cover forecast presented on the relevant Internet site, the site is applicable to any location.

The forecast provided has an hourly resolution and it is effective for various periods, ranging from 24 hours (forecast for today) up to 90 hours (forecast for the next three days). Verifications, performed through statistical processing, have enabled evaluating the degree of forecast reliability for today at 00: 00 a.m.

The result shows that forecasts performed by the ENEA system are completely in accordance with forecasts using other models which, however, do hot have the same operative applicability.

Updating the forecast every three hours enhances its reliability. Further forecast reliability enhancement can be obtained if the aerosol forecasts are available in the real time.

The following figures show the graphs of the forecast radiation prepared in the ENEA forecasting system for the site Casaccia (Rome) in a manner they appear every day on the ENEA site www.Solaritaly.enea.it, valid today and the following three days. It should not be forgotten that the solar radiation forecasts are used as an input in power generating models of the solar facilities, meaning that the most relevant inputs, like "revenues", into financial models and cash flows are coming from this forecasting.

2.3 Mathematical functions and physical laws

Latitude (ϕ):

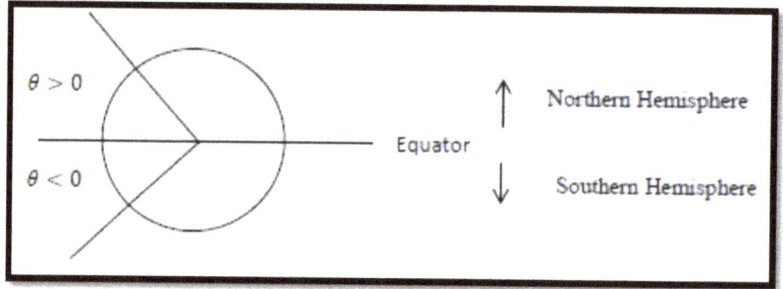

Picture 2.g - Latitude for northern and southern hemisphere

Day number (n): n=1 for January 1 & n = 365 for December 31

Declination (δ): $23.45° \leq \delta \leq 23.45°$

$\delta = 23.45°$ => Summer solstice => June 21, 2013

Longest day of the year

$\delta = 0°$ => Equinox => March and September Equinox

12 hours of day & night everywhere on Earth

$\delta = -23.45°$ => Winter solstice, shortest day of the year

$$\delta = 23.45 * \sin\left(360 * \frac{284+n}{365}\right)$$

Hour angle (ω): Solar time (ST) expressed as angle

$\omega < 0$ => morning

$\omega = 0$ => solar noon

$\omega > 0$ => afternoon

$15°/hour$ => e.g. $\omega=60°$ => 4pm, $\omega= -90°$ => 6am

$\omega = (ST-12)*15°$ where ST is in hour & fraction of hour (not min.)

e.g. 0.5 is 30 min past the hour

Zenith angle (θz): $cos\theta z = cos\phi * cos\delta * cos\omega + sin\phi * sin\delta$

Solar Azimuth angle (γs) : $\gamma s < 0$ in morning

$\gamma s = 0$ at solar noon

$\gamma s > 0$ in the afternoon

$$\gamma s = sign(\omega)\left(cos^{-1}\left(\frac{cos\theta z * sin\phi - sin\delta}{sin\theta z * cos\phi}\right)\right)$$

Where $sign(\omega) = \begin{cases} 1 \ if \ \omega > 0 \\ -1 \ if \ \omega < 0 \end{cases}$

Angle of Incidence Eqn: θ = (time, location, surface orientation)

In most general form: $cos\theta = cos\theta z * cos\beta + sin\theta z * sin\beta * cos(\gamma s - \gamma)$

1) For a plane rotated about a horizontal east-west axis with continuous adjustment to minimize the angle of incidence:

$$cos\theta = (1 - cos^2\delta * sin^2\omega)^{\frac{1}{2}}$$

The slope of this surface is given by: $tan\beta = tan\theta z|cos\gamma s|$

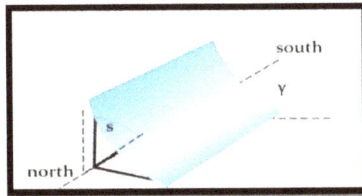

Picture 2.h - In picture s refers to θ

The surface azimuth angle for this orientation will vary between 0° and 180° if the solar azimuth angle passes through ±90. For either hemisphere,

$$\gamma = \begin{cases} 0° \text{ if } |\gamma s| < 90 \\ 180° \text{ if } |\gamma s| \geq 90 \end{cases}$$

2) For a plane rotated about a horizontal north-south axis with continuous adjustment to minimize the angle of incidence:

$$\cos \theta = (\cos^2\theta z - \cos^2\delta * \sin^2\omega)^{\frac{1}{2}}$$

This time the slope is given by:

$$\tan\beta = \tan\theta z |\cos(\gamma - \gamma s)|$$

The surface azimuth angle will be 90° or -90° depending on the sign for the solar azimuth angle:

$$\gamma = \begin{cases} 90° \text{ if } |\gamma s| > 0 \\ -90° \text{ if } |\gamma s| \leq 0 \end{cases}$$

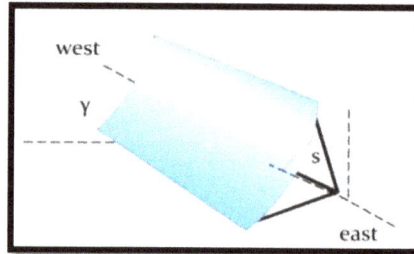

Picture 2.i - In this picture s refers to θ

3) For plane that is continuously tracking about two axes to minimize the angle of incidence

$$\cos\theta = 1$$
$$\beta = \theta$$
$$\gamma = \gamma s$$

4) For horizontal surfaces: $\theta = \theta z$

3. Solar calculations for energy yield

In this section we will take an example according best practice for usable thermal energy calculating and energy loss assessment of parabolic trough collectors, that collect heat.

Goal: quantify collector performance and sources of losses over 1-year.

Base collector performance model on SRCC Rating Sheet shall be considered.

➢ Nomenclature:

Q_u Useful heating (heat transfer to heat transfer fluid (HTF))

A_a Collector Aperture Area

T_m Collector Mean temperature in Kelvin

T_a Ambient Temperature in Kelvin

$F'(\tau\alpha)_{en}$ Collector Efficiency Factor (Normal)

$K(\Theta)$ Incident Angle Modifier (IAM)

U Wind Speed

G Irradiation on tilted surface ($G = G_{b,n}$ DNI $+ G_d(1+\cos\beta)/2$)

$$\frac{Q}{A_a} = F'(\tau a)K_{\Theta b}(\Theta)G_b + F'(\tau a)_{en} \, K_{\Theta d}\,(\Theta)G_d - c_1(T_m\text{-}T_a) - c_2(T_m\text{-}T_a)^2 - c_3 u \,(T_m\text{-}T_a)$$
$$+ \, c_4(E_L - \sigma T_a^4) - c_5\frac{dT}{dt} - c_6 uG$$

- Some of these terms can be ignored according to SRCC Rating Sheet ($K_{\Theta d}(\Theta)$, c_2, c_3, c_4 are zero).
- $\frac{dT}{dt}$ Term can be neglected since the HTF temperature is constant during the operation.
- As a 1st approximation, wind loss term $c_6 uG$ can be neglected for simplicity.

These approximations result in the following model:

$$\frac{Q_{coll}}{A_a} = F'\,\tau a)K_{\Theta b}(\Theta)I_{b,n}\, cos(\Theta) - c_1(T_m\text{-}T_a) = m_{htf}c_{p,htf}\,(T_{coll,e} - T_{coll,i})$$

Physically $K_{\Theta b} < 0$ is impossible and the Incidence Angle Modifier (IAM) is modeled as

$$K_{\Theta b} = max \; [1 - b_0 \, (\frac{1}{cos\theta} - 1), \, 0]$$

The total solar insulation striking the collector is

$$I_T = I_{b,n} \, cos\Theta + I_d \left(\frac{1+cos\beta}{2}\right)$$

The objective in this analysis is to identify the ultimate fate of IT in terms of how much is useful (for electricity production etc.) and how much is lost to the environment by various mechanisms.

When $IT > 0$ but is insufficient to drive Tm to 100^0 C (i.e., $Ta < Tm < 100^0$ C), the collector is described as being in a transient state and all the solar insulation is lost to the environment.

$$Q_{Loss,\ transient} = I_T$$

When $\dfrac{Q_{coll}}{A_a} > 0$ the system is operating and some fraction of IT is useful.

The remaining fraction of I_T is lost to the surroundings by the following mechanisms.
"Diffuse" losses:

$$\frac{Q_{L,d}}{A_a} = I_d \left(\frac{1+cos\beta}{2}\right)$$

Normal Optical Losses:

$$\frac{Q_{L,n}}{A_a} = [1\text{-}F' \, (\tau a)_{en}] \, I_{b,T}$$

Incidence Angle Modifier Losses:

$$\frac{Q_{L,\theta}}{A_a} = (1\text{-}K_{\Theta b}) \, F' \, (\tau a)_{en} \, I_{b,T}$$

Absorber Losses

$$\frac{Q_{L,A}}{A_a} = c_l \, (T_m - T_a)$$

Total insulation, which is sum of beam and diffuse insulation, must be equal to sum of all losses at any time.

$$I_T = I_{b,n} \, cos\Theta + I_d \left(\frac{1+cos\beta}{2}\right) = \frac{Q_{cool}}{A_a} + \frac{Q_{L,d}}{A_a} + \frac{Q_{l,n}}{A_a} + \frac{Q_{L,\theta}}{A_a} + \frac{Q_{L,A}}{A_a}$$

4. Solar technologies according using

Throughout history, various techniques of solar energy utilization have been developed, e.g. for cooking, metal smelting, even for military purposes (the Archimede's heat ray using an array of mirrors).

Anyway, the fossil fuels are most prevalent, and they are nothing but the solar energy accumulated by living organisms for millions of years, ready to use any moment.

The recent awareness of the relative scarcity of fossil fuels and the sensitivity to the environment pollution have imposed the need to use solar energy for purposes of technology in order to, at least partially, replace fossil fuels.

There are numerous technologies currently available and under development, and their classification, which is not simple, will probably soon become obsolete.

An attempt at classification is schematically shown later on, which includes relevant information on the typical areas of application and the degree of development of the major technologies, although each group reflects different situations, as will be shown in the chapters to follow.

The first group comprises technologies related to architecture, such as passive solar energy and daylighting.

Passive solar energy uses parts of buildings (walls, windows, roofs) to collect, accumulate and distribute solar energy for heating, cooling and ventilating the spaces in a building without the use of mechanical systems (ventilators, pumps etc...), as well as for integration with traditional air conditioning systems.

Daylighting uses the sunlight inside a building to reduce electric energy consumption, by letting the light in from the top (top lighting), from the side walls (side lighting) or using channels to direct light (core lighting).

Solar heat energy includes technologies for obtaining heat at low and medium temperatures for sanitary hot water, heating rooms in winter, cooking (solar ovens), water treatment (desalination) and some industrial processes (drying, maturing etc.).

A novel, interesting technology from this group is solar cooling of buildings in summer.

The technology of solar "power" is focused on producing electric energy in a direct manner (photovoltaic, thermoelectric using the Seebeck effect) or collecting the heat at high temperatures and its converting using thermodynamic cycles (Rankine, Stirling).

The solar chemistry includes various technologies to utilize solar energy as a driving force for chemical process reactions producing fuels, in purification treatments and other industrial processes.

Table 4.a – Classification of solar technology

	Technology	Sector of utilization	Development stage
	Passive solar energy	Residential	Developed
Solar energy in architecture	Natural lighting	Residential Offices Exhibitions	Demonstrated
	Sanitary water heating	Residential Hotels Sport facilities	Developed
	Winter air-conditioning	Residential Schools	Developed
Solar heat	Water treatment	Services	Demonstrated
	Cooking	Developing countries	Developed
	Thermal process	Industry	Demonstrated
	Summer air-conditioning	Residential Hotels Offices Trade	Demonstrated
	Photovoltaic	All	Developed
Solar power plants	Seebeck effect	to be defined	Research
	Concentrated solar power	Industry Services	Demonstrated
	Fuel production	Industry	Research
Solar chemistry	Depuration	Industry Services	Research
	Industrial processes	Industry	Research

5. Solar technologies according materials

Depending on the technologies used to produce the PV cell, different manufacturing process is taken place. Semi-conducting materials are the fundamental elements in making up a solar cell.

By different choice of semiconductors, crystalline silicon in a wafer form, thin films of other materials, and concentrated PV (CPV) are the technologies used.

PV cell technologies are usually classified into three generations, depending on the basic material used and the level of commercial maturity:

❖ **First-generation PV systems** (fully commercial) use the wafer-based crystalline silicon (c-Si) technology, either single crystalline (sc-Si) or multi-crystalline (mc-Si).

❖ **Second-generation PV systems** (early market deployment) are based on thin-film PV technologies and generally include three main families:
1) amorphous (a-Si) and micromorph silicon (a-Si/μc-Si);
2) Cadmium-Telluride (CdTe), and
3) Copper-Indium-Selenide (CIS) and Copper-Indium-Gallium-Diselenide (CIGS).

❖ **Third-generation PV systems** include technologies, such as concentrating PV (CPV) and organic PV cells that are still under demonstration or have not yet been widely commercialized, as well as novel concepts under development.

6. Photovoltaic energy: from manufacturing to operations

6.1) Crystalline silicon (c-Si)

As well known, from technology side the current mainstream of global operating plants is based on c-Si PV solution.

Technology of cells and modules based on crystalline silicon, first generation technology, still widely used and widespread in the market.

This technology generally uses two types of this semiconductor:

❖ mono-crystalline silicon
❖ poly-crystalline silicon, which shows lower costs and lower performances, if compared to mono-crystalline methods.

Other approaches, such as 'silicon ribbon' account for a negligible percentage in the market.

Picture 6.a Crystalline Silicon Cells

Polycrystalline is composed of a number of smaller silicon crystals. The multiple crystals create boundaries for electrons resulting in less efficiency comparing to mono-crystalline silicon.

However, polycrystalline can be produced at a lower cost than the mono-crystalline and it is used most in the solar industry.

Crystalline silicon solar cells are wafer based technologies which are considered the first generation solar cells.

Wafer based silicon solar cells have reached more than 24% cell efficiency using different metallization and doping designs.

This type of solar cells dominate the market with their low costs and the best commercially available efficiency.

They are a relatively mature PV technology, with a wide range of well-established manufacturers.

Although very significant cost reductions occurred in recent years, the costs of the basic materials are relatively high and it is also quite clear that further cost reductions will be sufficient to achieve full economic competitiveness: that should mean reaching the state of "grid parity", above all, in the wholesale power generation market in areas with modest solar resources.

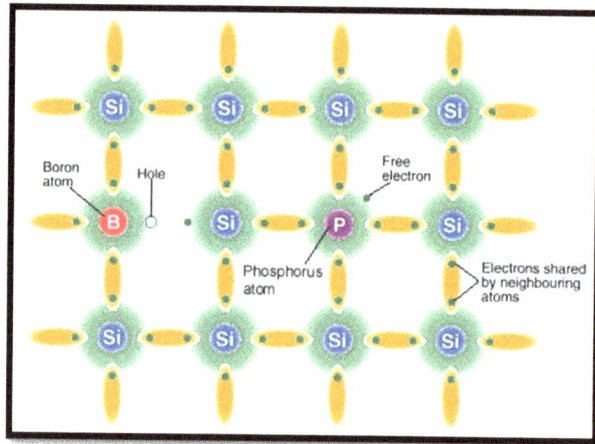

Picture 6.b - Doped Silicon

6.2) Basic structure of crystalline solar cell

To produce a solar cell, the semiconductor has to be doped.

"Doping" can be done either with positive charge carriers (p-type material) or negative charge carriers (n-type material).

Those two ties form a p-n junction close to the front surface. A p type crystalline wafer is doped with group V elements (commonly phosphorous) and n-type is doped with group III elements (usually boron).

In a non-doped single crystalline silicon lattice, which is called intrinsic Si, each Si atom shares four valance electrons with its neighboring Si atoms.

When the impurity atom with three or five valance electrons replaces a Si atom in the lattice, the electronic behavior of the crystal is changed. If the new impurity atom has five valance electrons, four valance electrons will be shared with neighbors and one excess electron will be free to move in the lattice.

This type of material, where the majority charge carriers are electrons, is called n type material. If the impurity atom has three valance electrons, there will be one electron missing in the covalent bonds of the Si atoms.

This unoccupied state of electron is called a hole and this type of material where the charge carriers are holes are called p type materials.

When n-type and a p-type material are brought together, one electron will flow from n-type material to p-type material and one hole will flow from p-type material to n-type material due to concentration gradient.

However, since both regions were electrically neutral initially, the electrons passing to the p-type material leave positive ions behind in the n region and holes passing to the n type material will leave negative ions in the p region.

Depletion region is formed at the junction plane which is populated with the electric dipole.

Therefore there will be a built-in electric field inside the p-n boundary pointing, from n region to p region.

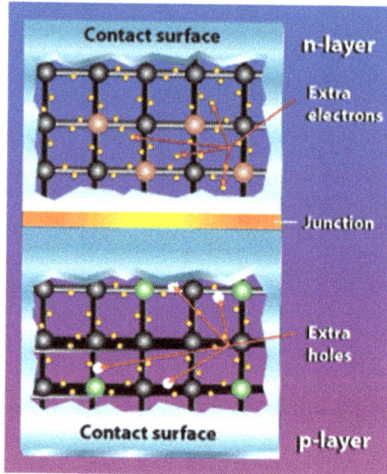

Picture 6.c - P/N junction

Metal contacts are attached to the top and the bottom surface, in order to extract the current. The front surface contacts are in grid formation, to enable light absorption in the p-n junction. The back side is fully metalized for the better charge collection. The anti reflective coating enhances light absorption by decreasing the reflectance of the front side.

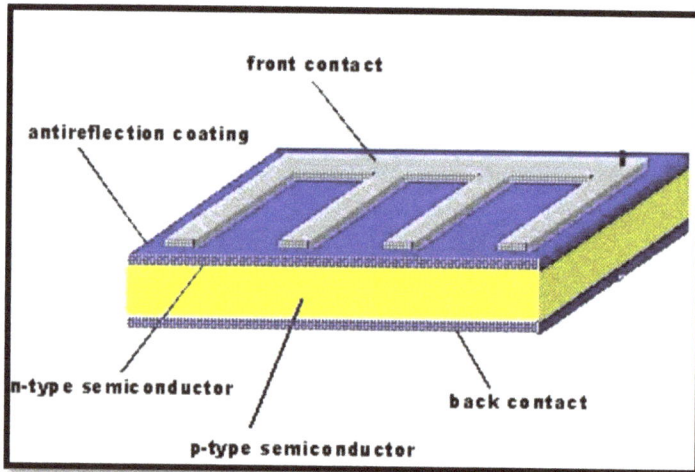

Picture 6.d - Basic structure of crystalline solar cell

6.3) Manufacturing of crystalline solar cell

Cell consists of one or two layers of a semi-conducting material. When light shines on the cell it creates an electric field across the layers, causing electricity to flow.

The greater the intensity of the light, the greater the flow of electricity is.

As we saw before, the most common semi conductor material used in photovoltaic cells is silicon, to very high degree of purity!

PV cells are generally made either from crystalline silicon, sliced from ingots or castings, from grown ribbons or thin film, deposited in thin layers on a low-cost backing.

The performance of a solar cell is measured in terms of its efficiency at turning sunlight into electricity.

A typical commercial solar cell has an efficiency of 15% about one-sixth of the sunlight striking the cell generates electricity.

Improving solar cell efficiencies, while holding down the cost per cell, is an important goal of the PV industry and that was the main trend during last six years.

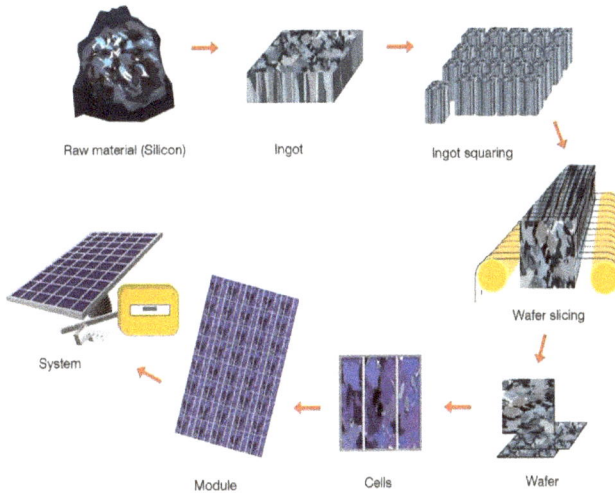

Raw material (Silicon)   Ingot   Ingot squaring

Wafer slicing

System   Module   Cells   Wafer

While single crystalline silicon solar cells posses higher efficiencies compared to multi-crystalline silicon solar cells, the cost of producing single crystalline wafers is much higher than producing multi-crystalline wafers.

The whole process of manufacturing can be divided into several steps:

- ➤ Chemical cleaning the starting wafers/pre diffusion cleaning;
- ➤ Etching wafers surfaces and Solid State Diffusion;
- ➤ Metallization;
- ➤ Anti-reflective coating.

Solar cells are devices that use photons to create electron hole pairs for photovoltaic conversion. Consequently, increasing the number of photons absorbed by the cell will increase the power output of the solar cell which is desirable for higher electricity generation. Silicon has a band gap of 1.12 eV implying that the useful spectral range for photovoltaic conversion is between 300-1100nm , as shown in Figure 6.e.

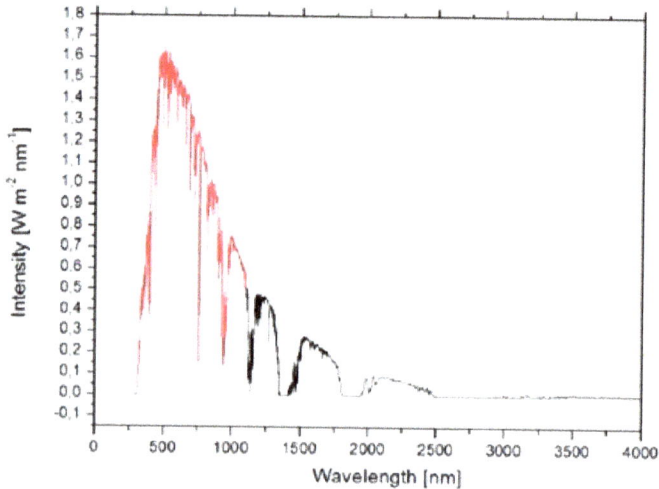

Figure 6.e - AM 1.5 spectrum at global tilt

For a 90% absorption of the incoming intensity, the 300-1000nm portion of the spectral range is absorbed in a thickness of 400μm; however for a full absorption, the solar cell will have to be at a thickness of more than 2000μm.

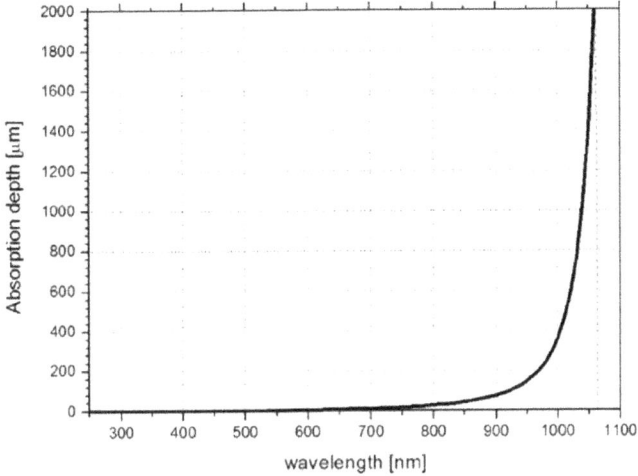

Figure 6.f - Absorption depth of Si for 90% absorption

However, in a single crystalline Si solar cell, the generated electron-hole pairs are separated at the p-n junction, who is close to surface and the pairs generated more than one diffusion length away from the junction has a very small chance to be collected by the external circuit.

Therefore, photons absorbed more than one diffusion length away from the junction will play no role on photovoltaic conversion.

For silicon the minority carrier diffusion lengths are given as in Table 6.g.

	n type	p type
Minority Carrier Diffusion Length	14 μm	140 μm

Table 6.g - Minority carrier diffusion lengths in crystalline Si

47

6.4) Pre-diffusion cleaning

Before the solid state diffusion step, the wafers must pass through a cleaning cycle in order to get rid of all organic-inorganic contaminants and the natural oxide that may have grown on the wafer surface. For this purpose the so-called RCA cleaning steps are followed for all samples before the doping process.

The cleaning procedure consisted of four main parts:

- *Natural oxide removal:* In this step, the wafers are dipped into a 1:4 HF:DI Water solution at room temperature for about 30 seconds. The process stopped when the surface of the wafer became fully hydrophobic.

- *RCA 1:* After rinsing of the wafers with DI water, the wafers are dipped into a solution of H_2O_2: NH_4OH: DI with a concentration 1.5:1:6 at 70°C for seven minutes. In this step, organic contamination was removed from the wafer surface.

- *RCA 2:* After rinsing of the wafers, the last RCA step which removed heavy metal contaminants from the surface is applied. For this, the wafers are dipped into a 1.5:1:6 concentrations H_2O_2: HCL: DI solution at 70°C for 7 minutes and then rinsed at DI water again.

- *Natural oxide removal:* After the RCA 2 process, another oxide layer grows on the surface due to H_2O_2. This oxide is removed in a 1:4 HF: DI water solution at room temperature for 30 seconds as in the first oxide removal process.

At the end of the cleaning process, the wafers were rinsed in DI water and, then, dried in a centrifugal drier and became ready for solid state diffusion.

6.5) Etching wafer's surfaces and solid state diffusion

Before the Solid State Diffusion, the etching wafer's surfaces is done by lithography. The next major step of the process is the diffusion of the cell junction.

The goal of this step is the formation of the p-n junction across the wafer so that created electron-hole pairs can be separated to generate the current.

Way to create a p-n junction is to dope the surface of an n type wafer with a group III (Boron) elements or alternatively to dope p type material with a group V (Phosphorous) element.

During the doping process, the dopant atoms already residing in the crystal is compensated with the same number of opposite type dopant atoms introduced during the doping process.

When the number of new dopants exceeds the native dopants in the surface region the type of the surface region is converted and two regions with different types of doping are created.

Crystalline Si is doped at high temperature (850 °C), for 30 minutes and the doping atoms diffuse into the material due to the concentration gradient.

After the diffusion process, RCA cleaning is used again.

6.6) Metallization

Metal contacts are applied to the top and bottom surfaces to extract the current. The top surface requires a fine-line metal grid to maximize the surface area that is exposed to the Sun, while back side is fully covered.

The contacts must be very thin (at least in the front) so that it does not block the sunlight to the cell.

There are several ways of metallization used for solar cells such as, thermal evaporation, e-beam evaporation, screen printing etc…

Although the screen printing technique is generally preferred for large applications vacuum evaporation and lithography are used for research purposes.

6.7) Anti-reflective coating

Pure silicon is very shiny and because of that it can reflect up to 35 percent of the sunlight.

The number of photons absorbed is a key factor that determines the current output of a solar cell.

An increase in the number of photons absorbed increases the current output and vice versa. It is then desirable to reduce the reflection from the cell surface and direct the incoming photons into the device active region.

To increase the number of photons absorbed by the solar cell, anti-reflective (AR) coating thin films are usually applied to the surfaces of the solar cells.

These coatings decrease the reflected intensity nearly to zero, around at some specific wavelengths and have a great effect on solar cell performance.

6.8) Operating principle

A photovoltaic (PV) system is a system converting solar energy directly into electricity and then its operation is based on the photovoltaic effect of the solar cells as shown here below.

PV technology started its development in the 1950s, with the first cell made of crystalline silicon produced at Bell Telephone Laboratories.

In 1958, it finds its first application in space (Vanguard I), while its terrestrial applications started in the mid 1970s, amidst growing research and development programs.

Since then the cost has gradually decreased, reduced by, among other things, the recent incentive policies that favored massive production of photovoltaic cells and modules, at costs lower than ever before.

Among various types of renewable energy sources, the PV system is, owing to its intrinsic characteristics, the most attractive and promising option with respect to the goal of grid parity of the photovoltaic energy and energy from fossil (mineral) fuel, achieved at the current prices in various geographic areas.

The advantages of the PV system that deserve special mention are modularity, the absence of the moving parts and fuel consumption, easy maintenance, simple operation, reliability and useful life of at least 20 years guaranteed by all manufacturers and installers (direct testimony of system longevity is that the historical, experimental systems are still in operation).

The operating principle of a solar cell is the photovoltaic effect, discovered by Becquerel in 1839; essentially, Becquerel observed that when the light illuminates certain material, such as silicon which belongs to the chemical group of semiconductors.

If the light has enough energy, it can knock the electrons free from their atoms.

With appropriate technologies, the solar cell creates voltage, also called potential barrier, acting on the above mentioned electrons to force their flow, creating a photoelectric current. The potential that generates the power is defined as photo-voltage.

Besides, it should be mentioned that each electron released corresponds, at least in case of silicon, to a chemical bond that lost an electron, designated as a hole.

Since free electrons may saturate various bonds, during their flow through a crystalline structure, this phenomenon generates an equal current of an opposite charge, consisting of holes that become free from time to time: such as phenomenon is designated the hole current, like the following graphic shows.

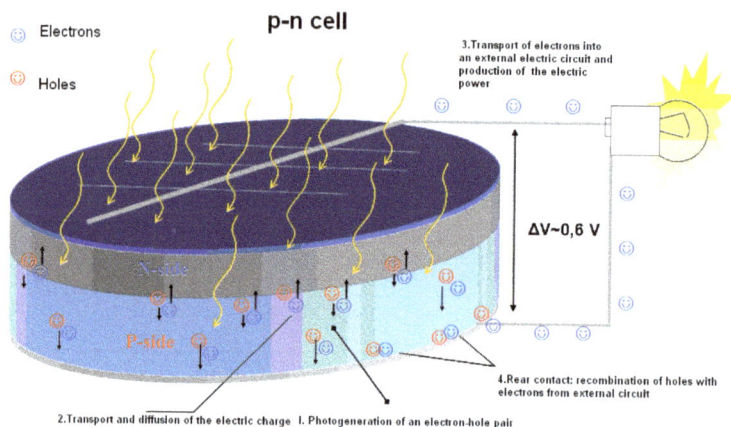

p-n cell

Electrons

Holes

3.Transport of electrons into an external electric circuit and production of the electric power

$\Delta V \sim 0,6$ V

P-side

4.Rear contact: recombination of holes with electrons from external circuit

2.Transport and diffusion of the electric charge 1. Photogeneration of an electron-hole pair

The electric current in a solar cell is actually the sum of the electron current and the hole current – both are forced by the potential barrier.

To better explain the operating mechanisms of charge and transport in p-n cells, we need to underline the crucial function of the potential barrier in a solar cell, because it stimulates the immediate separation of electrons and holes, preventing their recombination and forcing more electrons towards one of the sides of the cell and more holes towards the opposite side: in this way, the electric current in the external circuit goes to be established.

The so-called doping process is used for practical applications and to create the potential barrier. It is well known that silicon has 4 valence electrons which determine its chemical properties.

For example: if in a silicon block certain silicon atoms are replaced by phosphorus, which has 5 valence electrons and a part of silicon is replaced with boron, that has 3 valence electrons, then the material will establish a stable flow of charges, compensating for the excess of unsaturated bonds in the boron-doped part, by the excess of the electrons from phosphorus atoms.

The charges flows are a result of the electric field generated by the excess of opposite charges at different points of the material.

As the field is gradually canceled as a result of the flow of opposite charges that recombine, an equal electric field is generated but opposite in sign to the initial one. This is the potential barrier which determines the properties of a cell. In that specific case of silicon, its value is 0,7 volts.

6.9) Solar modules

To better focus on solar modules, we see that almost 90% of the solar modules in the market today are manufactured using technologies derived from electronics industry; even the basic active material such as crystalline silicon is the same, although the PV industry uses silicon of a slightly lower purity.

The major advantage of this material is its stability over time and strength, which allow processing using flexible technologies, resulting in manufacturing numerous electronic devices in the past years.

In the case of PV industry, the most recent industrial production processes make use of c-Si substrates in a square shape, sometimes with rounded edges (Fig. 6.h), with a side measuring 156 mm and thickness ranging from 0.2 mm, as we saw before, initially doped with boron, usually: p-type silicon.

The thin "tile" forms a junction through the doping process that produces a region of opposite conductivity (n-type), spreading over its entire surface, but penetrating the tile itself for less than 0.05 mm.

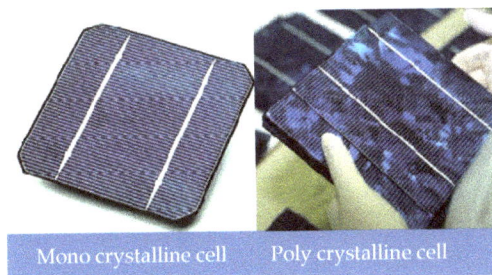

Mono crystalline cell   Poly crystalline cell

Figure 6.h – Crystalline silicon cells

The junction forms a potential barrier forcing the flow of electrical charges generated by light. To make a distinction, crystalline silicon is usually called "mono crystalline silicon".

A solar cell, then a solar module, is completed when an electric contact is created on both surfaces.

The contact on the light-exposed surface, called "front surface" hinders penetration of light into active materials, as little as possible and thus it generally has a form of a thin metallic grid of a total surface of just several cm^2, made via silk screen printing, using a paste based on silver as a printing paste.

The rear side contact is less problematic and it is made by depositing a thin layer of aluminum.

To minimize the losses due to reflection of light, the front surface is coated by a thin anti-reflective layer, made e.g. of SiO_2 .

Today, a well-made c-Si cell can generate the power of up to 9 A, at a voltage ranging from $0.6 - 0.7$ V and output power around 6.5 W in standard sunlight conditions.

To make substrate, the PV industry often uses poly crystalline[4] silicon instead of crystalline silicon.

This type of silicon differs from the former because it is apparently less homogenous: is consists of crystalline silicon grains of various structures.

The result is a substrate that keeps silicon's stability, albeit the electric properties and strength are slightly reduced, as well as its cost.

A poly-crystalline silicon cell can generate electric current up to 8 A at the standard sunlight, with a voltage ranging from $0.6 - 0.7$ V and an output power up to 5.5 W.

Solar cells constitute an intermediate product in the PV industry, supplying limited values of voltage and electric current compared to those normally required by user appliances; they are fragile, without electrical insulation and without any mechanical support.

Therefore they need to be assembled to obtain an integrated structure: a PV module.

The module is, in fact, the basic component of a PV system, with a strong structure easy to handle and not too heavy (a module weighs around 10 kg for each 100 W$_p$ on average) guaranteeing many years of operation even in difficult environmental conditions.

The solar module manufacturing process has various phases: electrical connection, encapsulation, mounting of the frame and junction box.

Electrical connection consists of connecting single cells in series or in parallel, in order to obtain the desired values of voltage and electric current; in order to reduce losses due to electrical disconnection it is required that cells of a single module have the same electrical properties.

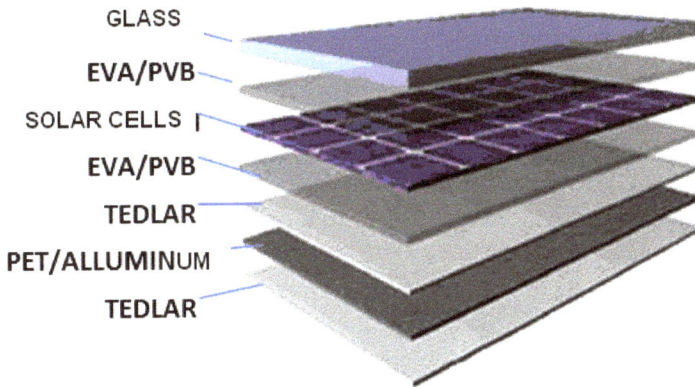

GLASS
EVA/PVB
SOLAR CELLS
EVA/PVB
TEDLAR
PET/ALLUMINUM
TEDLAR

Figure 6.i – Photovoltaic module with crystalline silicon, EVA/PVB, solar cells, Tedlar, PET/aluminum, Tedlar.

The encapsulation consists of incorporating the PV cell strings between the toughened glass panel (the surface exposed to the sunlight) and a rear panel made of glass, or more often, strong plastic material, in a hot lamination process using polymer glue.

In addition to protecting from weather conditions, encapsulation, importantly, needs to be transparent for the sunlight, stable to UV rays and temperature fluctuation, able to keep the cell temperature low.

Generally speaking, a life of a solar cell is infinite from technical side; hence, a module's life is determined by its encapsulation, currently estimated at 25-30 years.

To be underlined that for bankability and financial stakeholders side, that means business valuation and financial models (ex.: DCF method), we shall consider 20 years period, as well as a terminal value equal to zero of all the PV plant facility !

Aluminum frame mounting provides the module with higher strength, enabling anchoring to support structures.

The most widespread type of module in the market today, around 1-2 m² in size, has 36 to 72 cells of mono- or poly crystalline silicon cells.

Cells are arranged in parallel rows, while cells from one or more rows are connected in series.

A solar module produced in such a way has power that can reach above 300 W_p, and, depending on the cell's type and efficiency, it can reach operating voltage of around 15-20 V with electrical current ranging from 3 to 12 A.

Modules normally used for commercial applications have a total efficiency of 14-18% as range.

Their price varies considerably, but it can be said that today a poly crystalline module has a price ranging from 0.6-0.7 €/W_p, while mono crystalline module costs 10-20% more.

Table 5.12 provides the best results currently obtainable, through using solar modules and c-Si cells and poly-crystalline silicon cells (poly-Si).

Table 6.1 – Best characteristics of c-Si modules

Technology	Record cell efficiency (%)	Record module efficiency (%)	Guarantee (years)	Module surface area (m²)
c-Si	25	21.4	25	1.58
poly-Si	20.4	18.5	25	1.47

7. Solar plant configurations

The most advanced state of the art, as installed solar power plant, shows us the following practices of available configurations according grid connection.

Generally, they are two classes of PV systems:

1. Off-grid DC/AC PV system: DC without inverter, AC with inverter
2. Grid-tied system: with battery backup or no battery backup

depending on their connection to the utility grid.

In the first case, we need to refer to *Stand-alone* systems, where the generated electricity is directly supplied to an electric load and the excess is generally stored in batteries of the electric storage systems, making the electricity available for use in the hours of the day without insulation.

7.1) Off-Grid systems

✦ *Off-Grid DC system (without inverter):* the DC output is immediately directed to DC loads. Excess power is stored in the battery banks controlled by the charge controller. Common applications of this system are found in RVs, boats, cabins, farm appliances or rural telecommunication services. A backup generator may be included.

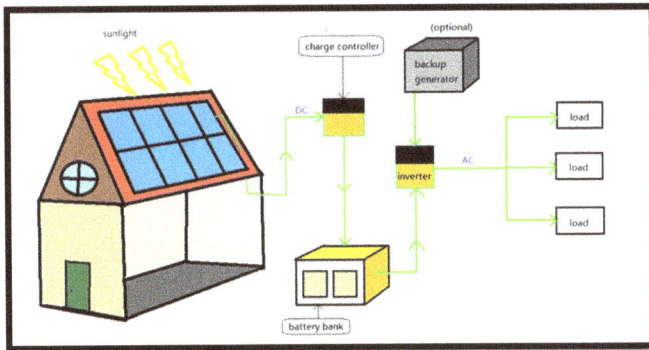

Picture 7.a - Off-Grid DC system (without inverter)

✦ *Hybrid system:* in this system, another renewable energy generator is added to generate more power. For example, the wind turbine or a geothermal power plant can be added to generate electricity, obviously from wind and geo sources as well.

This system is useful in places where the weather is sunny and, for example, windless during summer but cloudy and windy during winter.

This system is typically off-grid and the excess energy is stored in batteries. If neither the PV panel nor the wind turbine generates enough electricity, backup power such as a diesel generator can be added to generate the more energy.

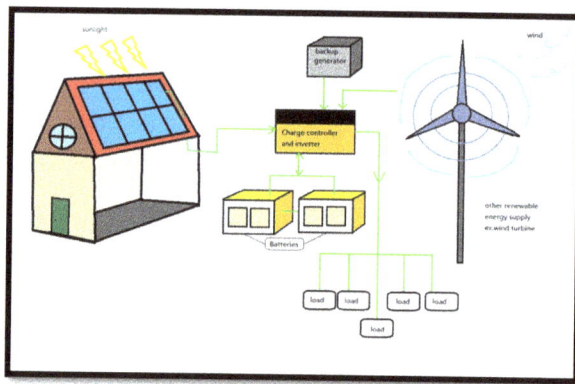

Picture 7.b - Hybrid System

A PV *stand-alone* system is advantageous in the cases when power has to be supplied to places out of reach of the electric power grid or in places where connecting to the grid would require investments too high, relative to expected consumption on the site.

Even without a favorable economic balance, a stand-alone PV system compared to an engine-generator for example, provides an opportunity to avoid a range of drawbacks, including fuel supply, noise, pollution, not to mention the PV system maintenance costs, which are much lower.

As small-scale of PV systems, for wide residential market, they bring huge added value in the social context, considering the solutions they may provide for developing countries, where 1.4 billion people have no connection to the power grid !

Compared to traditional sources, photovoltaics can easily be managed independently by local people and can be widely used, without building large distribution networks, making them economical and compatible with ecological environments that are not polluted by industrial activities yet.

Additional examples or fields of application of stand-alone systems:

❖ Water pumping, especially in agriculture.
❖ Supplying power to radio-repeaters for telecommunications.
❖ Charging batteries for civil and military use.
❖ Signals for traffic safety and for civil protection services.
❖ Supplying power to cooling devices for storage vaccines and blood in remote places.
❖ Supplying power to households, schools, hospitals, shelters, farms, laboratories in remote villages.
❖ Water treatment and purification.
❖ Cathodic protection in industry and in the oil and metal sectors.

These systems require electric storage systems (typically batteries) to ensure the energy supply even at night, and in unfavorable weather conditions.

7.2) Grid-tied systems

In the second case, we are talking about the *Grid-connected* systems, where the electric energy is converted into alternating current, that supplies the load (user) and/or supplied to the grid as well, with which it operates in the net-metering mode.

➕ *Grid-tied system without battery backup:* in this system, the generated DC is converted to AC and used on-site.

The solar power production is monitored by the solar production meter. If there is an excess energy, the energy can be fed into the electricity grid. If the PV system does not generate enough power because of higher demand, needed energy can be drawn from the grid.

This process of drawing or feeding electricity to the grid is monitored by the export/import meter (SCADA systems).

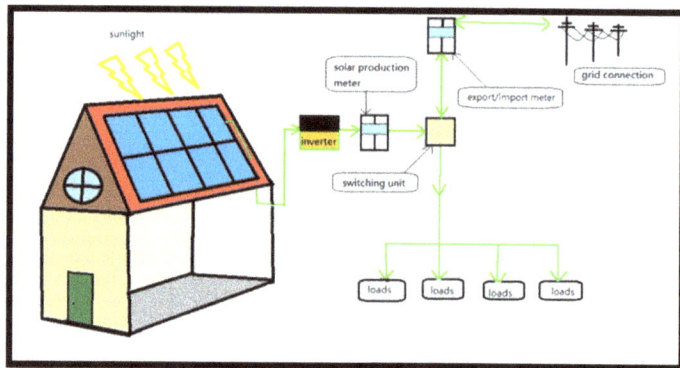

Picture 7.c - Grid-tied system without battery backup

↓ _Grid-tied system with battery backup_: in this system, the converted AC is used on-site or stored in batteries.

The charge controller monitors the battery capacity and excess energy is stored in the batteries for backup.

If the batteries reach their full capacity, the excess energy can be fed into the electricity grid.

On the other hand, if the PV system does not generate enough power, needed energy can be drawn from the grid. This process is done automatically, through a net metering program.

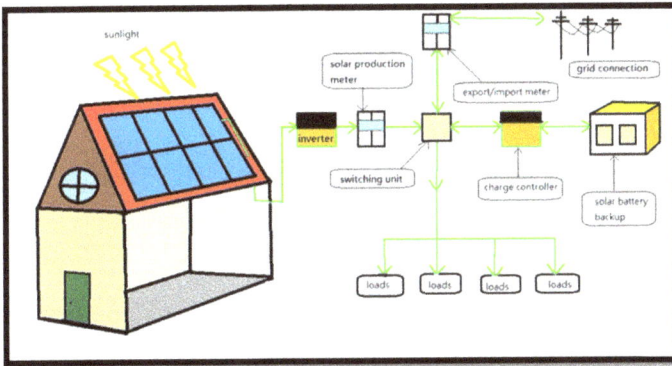

Picture 7.d - Grid-tied with battery backup system

Finally, grid-connected systems are used in places with available and sustainable electrical connection to the national or distribution grids, with the aim of generating electric power in a manner competitive to conventional sources, reducing the environmental impact of the electric power production and, in almost all industrialized countries, contributing to environmental objectives related to the reduction of the CO_2 emissions.

Typical applications range from small-scale systems in residential sector (< 20 kW), a segment characterized by systems integrated into residential buildings and connected to the low voltage (LV) distribution network, to medium-large scale systems (with power ranging from 20-200kW to 200-1000 kW), to photovoltaic power plants (>1000 kW).

In Italy, for example, PV systems with generated power higher than 100 kW and below 6 MW are connected to the medium voltage (MV), meaning under distribution networks; above 6 MW are connected to the high voltage (HV), under national grid.

Other applications are related to support for weak branches of the power grid or grid of small islands.

In general, grid-connected systems do not require the use of electric storage systems; however, with an increased share of renewable power generation, the new power grid requirements tend to give a central role to electric storage even in the grid-connected applications.

I like to underline that new frontier of global power generation, meaning the second green revolution, from business side as well as from social impact side, shall be strongly linked to the *smart city* market introduction, based on reaching mature technology and low cost application of batteries and/or energy storage systems.

Figure 7.e – Grid-connected PV plant

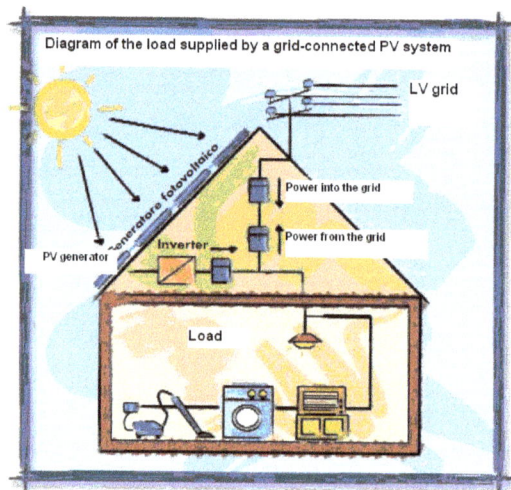

7.3) Components of Photovoltaic Systems

Schematically, a PV system consists of a chain of components with a task to capture the solar energy and convert it into electricity, with characteristics useful for the end user (Fig. 7.f).

Its principal parts are the PV generator, the system of electric power conversion and control and, depending on the type of system, the electric storage system.

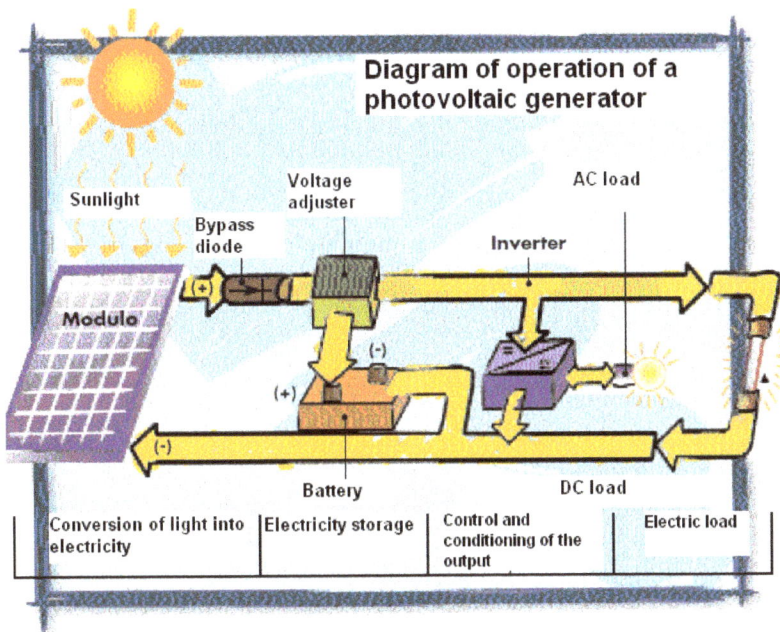

Diagram of operation of a photovoltaic generator

Sunlight
Voltage adjuster
AC load
Bypass diode
Inverter
Modulo
(-)
(+)
(-)
Battery
DC load

| Conversion of light into electricity | Electricity storage | Control and conditioning of the output | Electric load |

Figure 7.f – Components of a PV system

7.4) Photovoltaic Generator

The PV generator is a clearly 'visible' part of the system – it is the surface capturing the solar radiation whose dimensions are proportional to the total power output of the system.

The basic element of the PV generator is the module, a commercial component manufactured by assembling and encapsulation of PV cells.

An appropriate number of modules connected in series / parallel make up a generator or a PV field with the desired characteristics of current and operating voltage.

Its main electrical parameters are the nominal output, expressed in W or in peak W (W_p), i.e. the output delivered by the generator in the nominal (standard) conditions (irradiance of 1000 W/m^2 and the module temperature of 25°C) and nominal voltage, i.e. the voltage that delivers the nominal output.

The generator is usually subdivided into sections or strings. Each section, individually separable, is equipped with a bypass diode and an appropriate surge protector.

A PV generator has a support structure that is an important accessory component, designed to meet the requirements of low cost and high reliability.

The structure shall always be oriented and inclined to maximize the solar irradiance.

7.5) Electric Power Conversion, Control System, PV Inverter

Variable levels of voltage and current output of the PV generator, caused by the variability of the solar irradiance, is often inadequate for user's specifications, which often require alternating current to power the load directly or to connect to the grid with constant levels of the output voltage from the generator.

To shortly anticipate the main concept of inverter, we have to point out that this component is a fundamental device for output power conditioning and control, whose basic task is to convert DC generated in the PV field into AC (AC/DC power converter).

Operating as a DC transformer with a variable transformation ratio, the inverter provides a PV generator with a constant output voltage, regardless of the voltage fluctuations caused by varying insulation and cell temperature.

It is designated as 'the Maximum Power Point Tracker' (MPPT), and also acting on its operating point, it allows the PV generator to detect an optimal load at any given time, in order to deliver maximum available power.

If handled appropriately, such a device may perform the function of the control of battery charging.

In order to reduce costs, in small-scale devices, an AC/DC converter is replaced by a voltage adjuster calibrated to prevent overcharging the batteries.

Other built-in functions are related to the output power quality (filters) and security (blocking devices and interface protection).

To connect the system to the power grid in a typical case of net metering, it is necessary to have a counter for measuring the power drawn from the grid and a counter for the power delivered to the grid.

Moreover, a solar inverter operates to convert the variable direct current (DC) output of a photovoltaic (PV) solar panel into a utility frequency alternating current (AC), that can be fed into a commercial electrical grid or used by a local, off-grid electrical network.

In practice, as above mentioned, the static power converters used in photovoltaic systems allow converting the AC electric quantities from the generator into DC quantities, suitable for the transfer of energy to the grid or to consumers using AC.

Due to that's a critical component in a photovoltaic system for both: project execution and bankability key-condition, allowing the use of ordinary commercial appliances, it's important to go deeper into the inverter special focus, as follows.

DC/AC conversion is conducted via "conversion bridge", which uses semiconductor devices (generally IGBT, Insulated Gate Bipolar Transistor or MOSFET, Metal-Oxide Semiconductor Field Effect Transistor), driven by sequences of controlled command impulses.

The major part of commercial inverters conduct the bridge, switching at frequencies higher than the grid (several tens of thousand Hz), using the PWM (Pulse Width Modulation) technique, that modulates the impulse duration; the technique allows generating series of impulses whose duration is proportional to the value of the sine wave, assumed at the certain instant.

The converters are classified into two basic categories:

1. LCI or Line Commutated Inverter
2. SCI or Self Commutated Inverter.

In the LCI inverters, the grid voltage, necessarily active, constitutes the referent value for the generation of command impulses (switching on and off) of the semiconductor components.

On the other hand, the SCI impulses are generated over an appropriate control system with an autonomous clock, that establishes its referential frequency and with a power supply that allows switching and/or blocking power semiconductor devices.

In the case of grid-connected PV systems, DC to be converted into AC is generated in a PV field, while, in stand-alone systems with an electric storage, it is the power occurring in the nod that generates the PV field – electric storage system (classification of PV systems is given before).

Inverters for the grid-connected systems always have an MPPT, which allows the converter to adjust its own input impedance to achieve the impedance level required to transmit the maximum electric power to consumer.

This function is usually performed by the first bridge of the DC/AC conversion, through a control unit with a microprocessor.

In particular, voltage or current are regulated at the output (depending on the technology applied), in order to the inverter to be seen by the grid, in the first case, as a voltage generator that regulates its load angle (the difference between the generator and the grid) and to transmit maximum output, as well as, in the second case, as an electric current generator that delivers to the grid the current proportional to the maximum transmittable output.

The second stadium of the DC/AC conversion, synchronized with the grid frequency, provides that output power which has the desired characteristics of voltage and frequency.

In case it is not particularly required to adapt the input voltage (at the PV generator side) and at the output (at the load or grid side), the MPPT control and the output quantity adjustment (voltage and current) may be conducted through a single stadium of the DC/AC conversion.

The fig. 7.g below shows the flowchart of a converter suitable for grid-connection.

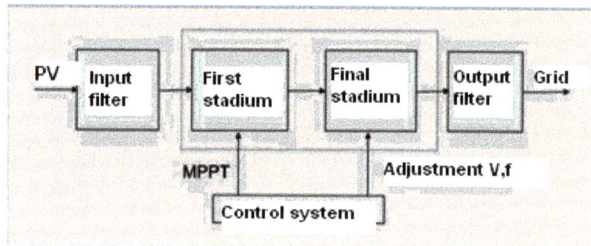

Figure 7.g - Flowchart showing the principle of DC/AC converter for grid connection

After the final stadium of conversion, there is yet another section for filtering harmonic frequencies of the power delivered to the grid and also devices for interface protection, on the load side as usually:

✓ devices of maximum and minimum voltage,
✓ maximum and minimum frequency
✓ maximum power

that can meet the specifications for grid connection (according national Grid Code).

Depending on the system architecture, inverters may have the transformer installed between the two conversion stadiums (High Frequency Transformer, HFTR), or alternatively at the output of the final stadium (Low Frequency Transformer, LFTR). The transformer allows adapting the converter's output voltage to the grid voltage, and providing conditions for metallic separation of the PV generation systems and consumers, with a possibility of a different way of managing the PV field.

The research and development efforts of the major subjects in the sector (public and private research centers, manufacturers etc.) have a goal to develop and manufacture inverters with ever higher performance indices, able to supply additional services to the grid.

The introduction of so-called smart inverters, that offer exactly this type of service, in Italy, is already planned by Fourth Energy Bill, whose implementation is urged by the new Decision of AEEG (Regulatory Body for Energy and Gas) 84/2012 (March, 2012), which introduces additional regulations and technical requirements for connecting PV systems to the LV and HV grid. The attention of the Bill is focused on electric energy quality and grid safety.

Smart converters perform several tasks regarding the grid. The main tasks iclude:

- to be insensitive to sudden changes of voltage;
- long-distance functions for voluntary separation off the grid;
- highly selective protection;
- supplying and absorbing the reactive energy;
- regulating the power output delivered to the grid to stabilize it;
- ensuring the grid is safe in the islanding conditions.

The above functions shall be complemented by the ability to communicate with the grid and with consumers, as required by ever higher number of smart grids, able to fully integrate the distributed power.

PV inverters have special functions adapted for use with photovoltaic arrays, including maximum power point tracking and anti-islanding protection.

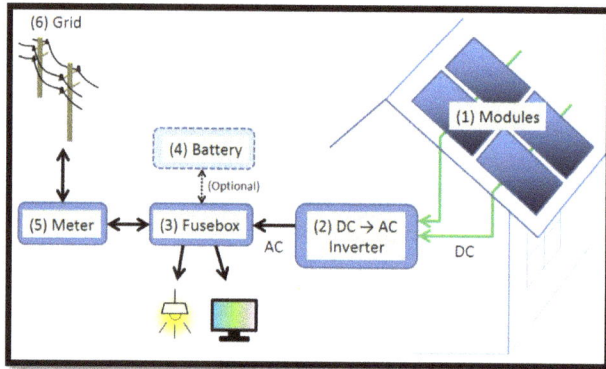

Picture 7.h - Simplified schematics of a grid-connected residential photovoltaic power system

Four major functions or features are common to all transformer-based, grid-tied inverters:
- Inversion
- Maximum power point tracking
- Grid disconnection
- Integration and packaging

Also, PV inverters may be classified into three broad types:

- *Stand-alone inverters*, used in isolated systems where the inverter draws its DC energy from batteries charged by photovoltaic arrays. Many stand-alone inverters also incorporate integral battery chargers to replenish the battery from an AC source, when available. Normally these do not interface in any way with the utility grid, and as such, are not required to have anti-islanding protection.

- **Grid-tie inverters**, which match phase with a utility-supplied sine wave. Grid-tie inverters are designed to shut down automatically upon loss of utility supply, for safety reasons. They do not provide backup power during utility outages.

- **Battery backup inverters**, are special inverters which are designed to draw energy from a battery, manage the battery charge via an onboard charger, and export excess energy to the utility grid. These inverters are capable of supplying AC energy to selected loads during a utility outage, and are required to have anti-islanding protection.

Picture 7.i - Inverter for grid connected PV. University of Mugla, Turkey

7.6) Electric storage systems

The storage of PV generated electricity may prove useful for performing major functions of the PV system itself, both in the case of stand-alone and grid-connected systems.

An electric storage facility allows time-shift, i.e. the use of energy even at times when it is not generated and it may be used to prevent grid-connected PV systems from losing the generated energy in case the power grid, to which the plant is connected, is interrupted, i.e. for anti- islanding.

Besides, a PV plant with an electric storage system is able to provide the grid with a more regular load profile and, therefore, to enhance integration of the system into the grid.

Even though there are various methods of storage of the electricity (hydraulic, compressed air CAES, flywheel, hydrogen etc.), by far the most utilized one is the electro-chemical method, i.e. family of batteries.

The most commonly used device is lead-acid batteries, due to their high level of technology and low price, but on the other hand, the use of lithium batteries is increasing (lithium ions), owing to its high efficiency and specific power.

In large-size PV systems, there is a notable use of sodium-sulfur and nickel-sodium batteries with high specific power and energy.

To better clarify and go deeper and deeper on this specific matter, I shall postpone later on such a topic to an additional book, focused on power storages and technologies of battery.

8. Thin film

Thin film is an alternative technology that may use less or no silicon in the manufacturing process.

Technology using thin-film solar photovoltaic modules, designated as second generation technology, with demonstrated maturity and industrial application.

The thin-film technologies, which share today a consolidated position in the market, include three different types of materials/semiconductors used in thin films:

➤ Amorphous/microcrystalline silicon (a-Si / μc-Si),
➤ Copper and Indium (di)selenide (CIS or CIGS in case it includes Gallium),
➤ Cadium-telluride (CdTe).

From the manufacturing side, the thin film PV cells are constructed by depositing extremely thin layers of the semi-conducting materials onto a low-cost backing such as: glass, stainless steel or plastic.

A conducting layer is then formed on the front electrical contact of the cell and a metal layer is formed on the rear contact.

Semiconductor thickness is several microns (0,001 mm) or even lower than a micron, in case of amorphous silicon.

Second-generation thin-film PV technologies are attractive because of their low material and manufacturing costs, but this has to be balanced by lower efficiencies than those obtained from first-generation technologies.

Thin-film family's market are less mature than first generation PV of course, then still have a modest market share, except for utility-scale systems.

They are struggling to compete with very low c-Si module prices and also face some issues of durability, bankability requirement (sometimes), materials availability, materials toxicity (in the case of Cadmium) as well.

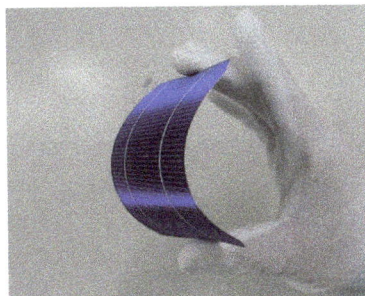

Picture 8.a - Thin film

8.1) Thin film modules

As anticipated, thin film technologies, so called second solar generation, start up few years ago from industrial requiring of cheaper materials and lower manufacturing costs, in order to face very high market potential (ex: taking over the global PV shortage) and while accepting the lower sunlight conversion efficiency, if one compares it with conventional PV c-Si modules.

In this way, R&D and commercial manufacturers focused on low-cost materials using: plastics, glass, metal as a support for very thin layers of semiconductors with higher sunlight exploitation capacity than c-Si.

The manufacturing technologies employed in these cells are suitable for plants with high automation and low labor intensity, with high production levels: a thin semiconductor film is deposited on a large area; appropriate laser cuts can divide the material into cells directly connected in series, making monolithic modules of up to 5 m^2 in size.

This process avoids producing large amounts of waste material that normally occur when cutting ingots for c-Si wafers, reducing the assembling price of modules and increasing their reliability at the same time: it eliminates the wafer positioning problem and there are only two solder junctions compared to tens of junctions in serial wafer connections in first generation module cells.

The temperature of the deposition process is relatively low (150-600 °C) and higher concentrations of impurities may be tolerated, resulting in reduced energy consumption and the relative payback time is short, i.e. the period the product needs to be exploited to generate the amount of energy consumed to manufacture it: around one year in South Europe, which has higher insulation and one year and a half in Central Europe.

The possibility to reduce the costs of thin film modules manufacturing and, thus, to provide products at more affordable prices allowed the thin-film technologies to gradually increase its share in the PV market - up to 21% of the total production in 2009.

In the last couple of years the situation has changed dramatically: the c-Si module prices have dropped rapidly and sharply, reducing and even canceling the competitive margin of the thin-film technology, as a result of:

○ Price reduction of poly crystalline silicon, sold at only 30 $/kg with forecasts of further decrease as low as 20$/kg, as a consequence of major investments in this sector and a sharp increase in production in the past years.

- Economies of scale lead to building giant plants capable of producing modules to generate several hundreds MW per year, located in low-cost labor countries (mainly China and Taiwan).
- Increasing the average module efficiency due to improvement of the production lines.
- And, above all, increased the gap between the production capacities (supply) and demand.

Manufacturers in the thin-film technologies have thus reacted very cautiously, preferring to postpone new major investments and in 2011, their market share decreased to 13% of 34.8 GW (in 2010, it was 16% of 19.9 GW).

The expectation of a new price decrease leads to the assumption that in the sector of thin-film PV system, only the largest companies or those integrated into economically strong groups will survive, as well as those able to come out with an integrated offer.

The best prices in the global market, recorded in March 2012, were 1.06 $/W (around 0.8 €/W), for c-Si technology and 0.84 $/W (0.63 €/W) for the thin-film technology[10], numbers very close to production costs.

It should be mentioned, however, that in absolute terms, the manufacture of thin film modules is steadily increasing, from 3181 MW in 2010 to 4575 MW in 2011, an increase of 44%.

The CdTe technology accounts for 45%, thin film of silicon 36%, while CIGS accounts for 19%.

The continuous vitality of the thin-film module sector can be understood if the following characteristics, that make these modules especially suitable for architectural integration on roofs and building facades, are taken into account.

- Their homogenous color (brown-black), linked to the fabricating technology by deposition, directly on glass, is highly appreciated from the esthetic point of view.
- To various degrees, they all have improved energetic efficiency, compared to c-Si modules of equivalent nominal power, as a result of the lower reduction rate of conversion efficiency at working temperatures (that in summer months may even be higher than 70 °C), making the module retroventilation problem, less prominent in architectural applications; in addition, they perform well in the conditions of diffuse light and in low levels of irradiance (with respect to the 1000 W/m^2 standard), making them less sensitive regarding ideal orientation, which is not always practicable in these applications.

The thin film technologies are therefore ideal for the manufacture of PV modules, suitable to make the new business model of distributed power generation, with devices that generate electric power integrated into buildings, where consumers live and work, so as to suit their needs.

8.2) Thin silicon film cells and modules

The simplest solar cell, made of amorphous silicon that can be manufactured, has three sequential layers:
- ✓ a very thin p-type layer (doped by boron),
- ✓ 5-20 nm thick and, almost transparent, one intrinsic (without doping),
- ✓ one of n-type (doped by phosphorus).

In the p-i-n structure, a positive charge region is created, corresponding to the i-n interface and p-i region of negative charge, as a result of the electron flow from the doped n-layer into the doped p-layer, forming an electric field E (Fig. 8.b).

The photons, that constitute solar radiation, first get absorbed by the intrinsic layer forming electron-hole pairs that get separated in the electric field and flow toward electrodes, connected to an external circuit.

The efficiency of sunlight conversion in these devices is initially higher than 10%, but the exposure to sunlight in the first hundred hours results in a sharp degradation that reduces their efficiency up to 30% (Staebler-Wronski effect).

The manufacturing process of amorphous hydrogenised silicon employs PECVD (Plasma Enhanced Chemical Vapor Deposition) technology and the raw material used is pure silane (SiH_4) gas or mixed with other gases.

In highly diluted mixtures with hydrogen (>95%), a microcrystalline silicon film is obtained, a mixed structure with a fine-grained structure of microcrystalline silicon in a matrix of amorphous silicon.

This material has recently gained particular attention since it has higher stability than amorphous silicon when exposed to sunlight.

The most common substrate is glass, with a front electrode on top of it, a transparent and conductive film consisting of a metallic oxide (TCO), typically SnO_2 or ZnO.

The rear electrode is fabricated as a bi-layer electrode with a high reflectance and a thin layer of ZnO and a metallic, silver or aluminum layer.

The use of flexible substrates of steel or polymer sheet allows fabrication of flexible products, easily adaptable to rounded surfaces, making them especially attractive in the building industry.

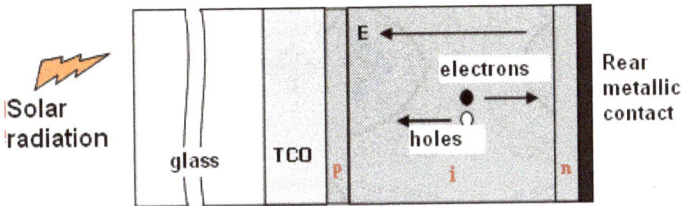

Figure 8.b – Diagram of a typical PV cell with a thin a-Si film

Solar cells with a thin film made of amorphous or mono crystalline silicon with a single p-i-n junction have stabilized, quite low conversion efficiency (after the first one hundred hours of sun irradiation), laboratory record is around 10%, compared to 7% in the case of industrial products.

In order to obtain better performances, multiple junction solar cells are produced, i.e. structures containing two or three junctions that are superimposed and, then, connected in series, each one made of intrinsic materials of high absorbing power and capacity to convert solar radiation into various determined wavelength intervals.

Double junction, also called micromorphous, is obtained with an a-Si front cell and a micro crystalline Si bottom cell.

The tandem cell is characterized by a notably smaller degradation, due to the use of micro crystalline Si, with a thickness of at least 2-3 μm in order to make the best use of solar specter.

Therefore, it is necessary to develop a high speed deposition process (>10 Å/s), a goal unachievable with large dimension substrates; moreover, as a matter of fact, commercial products with such structure are not larger than 1.5 m^2 in size, but their efficiency is close to 10%.

The lowest industrial cost is approximately 0.55 $/W, that also corresponds to large-scale module production with the simplest structure available (a single p-i-n junction) and lower efficiency.

The use of insulation substrate, such as glass, allows making monolithic thin-film modules (Fig. 8.c).

It is necessary to subdivide a device installed over a large surface into cells and to interconnect them in series in order to:

a. reduce electric power generated by a module exposed to sunlight, proportionate to the cell size and therefore to reduce energy the losses due to the Joule effect;

b. bring the output voltage to the values convenient for the intended purpose of module.

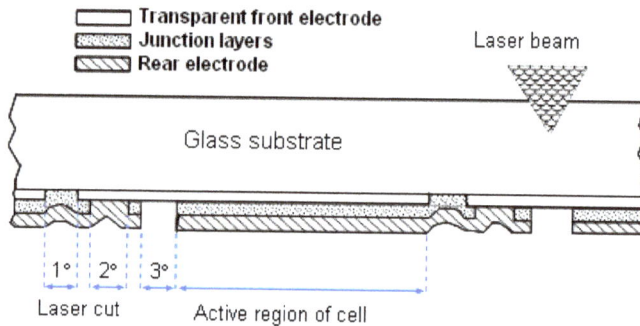

Figure 8.c – Diagram of the subdivision into interconnected cells in a thin-film module on glass substrate with the front electrode as the first layer.

This is achieved by selective cuts (Fig. 8.c) of various layers that constitute the PV device using laser of the appropriate wavelength (*laser scribing*).

The first cut is carried out on the contact deposited on the substrate; next, the layers that constitute the junction are deposited and subdivided with the second cut, immediately next to the first; at the end, the final contact is carried out, making the final cut next to the second one, leaving only the basic electrode intact.

This sequence allows dividing the module into cells, each 0.5-1 cm in width and length equal to the substrate dimension, with its first electrode connected along the longest dimension with the second electrode of the preceding cell, while its second electrode is connected to the first electrode of the subsequent cell.

Laser beam penetrates the glass if the first electrode is transparent (*back scribing*).

The cut region is not active and therefore its dimensions are minimal, compatible with the requirements of insulation and contact resistance reduction.

This technique, however, cannot be applied in case of metallic, non-insulating substrates, such as steel or aluminum sheets, when cells are first mechanically subdivided and subsequently interconnected using exactly the same method as in c-Si cells in wafer.

8.3) CdTe Thin Film

Cadmium telluride or CdTe is a photosensitive compound, where atoms of tellurium and cadmium are connected by a strong ionic bond, that gives the material high stability and prevents degradation caused by solar radiation.

Its optical properties, including high capacity of absorption and conversion of photons in the 250-825 nm intervals, where most of the solar radiation is concentrated, make it especially convenient for PV manufacturing.

CdTe is a p-type material, that couples with cadmium sulfide (CdS) in a very thin film, to keep its transparency, to form a junction and an electric field, that allows separation of electrical charges generated by absorption of sunlight in cadmium telluride (Figure 5.54).

In most of cases, the industrial manufacturing is carried out using PVD (Physical Vapor Deposition) technique.

CdTe is deposited in a layer 2-6 μm thick, and once deposited, it is thermally processed at 400 °C in the chlorine (Cl) atmosphere, essentially important for increasing the dimensions of crystalline grains and enhancing electronic properties of the material, as well as hence the performances of the final products.

It should be noted that the PV modules with CdTe thin film and their production techniques and disposal, at the end of its life, are the central subject of an ongoing debate related to cadmium's high toxicity.

There are also doubts related to the long-term perspective of investing into this technology, due to scarcity of tellurium, which occurs as a byproduct in the copper mining industry; anode mud collected in the electrolytic metal refining process yields barely around 100 g of tellurium per one ton of copper.

Nevertheless, the American company First Solar, which produces only CdTe modules was ranked second largest on the global level in 2011, producing 1.981 GW PV modules.

Their modules with 11.8% efficiency are the best commercially available product based on this technology, with production cost of 0.75 $/W.

First Solar created a cell with an 17.3% efficiency, a world record for small-scale PV modules.

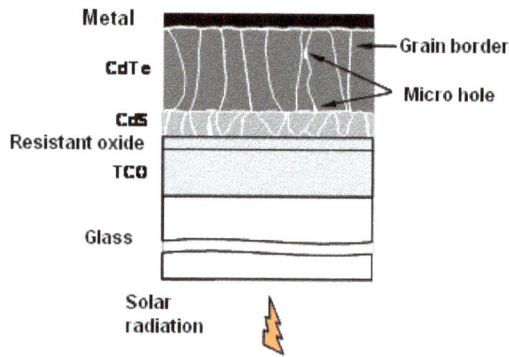

Figure 8.d – Diagram of a PV cell with CdTe thin film on glass.

8.4 CIGS Thin Film

Copper and indium diselenide is a poly-crystalline photosensitive p-type material, characterized by high stability, excellent absorption of sunlight in a wide range of wavelengths, from ultraviolet to infrared.

Its composition may vary with high tolerance between copper and indium; to change its response to solar specter indium is partially replaced by gallium or aluminum and selenium by sulfur.

Solar cells with a CIGS thin film (Fig. 8.e) display record efficiency compared to all other technologies of this type, reaching 20% on a small area by the American research institute NREL.

The best commercial product is made by MiaSolé with an efficiency of 14.5%, but there are other companies producing modules with efficiencies above 12%.

The industrial technique of CIGS film production has two stages:

1. deposition of Cu, In and Ga on the substrate covered by a molybdenum, using a low-cost process of high deposition uniformity (e.g. PVD or sputtering),
2. subsequent selenization, in a thermal process, at 400-600 °C in the atmosphere with Se.

CIGS is a p-type material and also, similar to CdTe, the junction is generally formed with a thin layer of cadmium sulfide (CdS).

There are also alternative solutions regarding CdS, to avoid the use of highly toxic cadmium in fabrication process, however, the results are poor.

The device is completed with the front transparent contact, a thin layer of a resistant oxide, like ZnO in this case, or highly conductive doped oxide: e.g. ZnO doped by boron or aluminum.

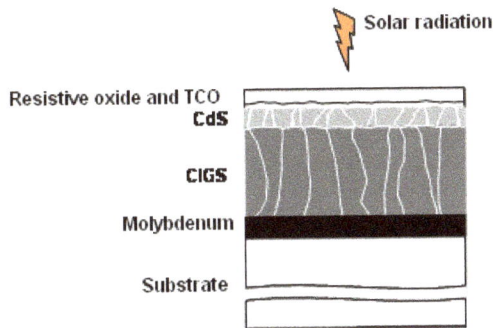

Figure 8.e – Diagram of a PV cell with CIGS thin layer.

9. CPV: concentrated PV cells

Concentration of sunlight onto photovoltaic cells is not a new idea. Certain experiences draw on the very beginning of the PV systems, but the actual advent of CPV technology is considered to have started in 1975 amid the oil crisis, at the initiative of American ERDA (Energy Research and Development Administration), nowadays DOE (Department of Energy), who financed a program on CPV, coordinated with Sandia National Laboratories.

This period marks the beginning of numerous industrial initiatives in this sector, by Motorola, RCA, GE, Martin Marietta, E-Systems (later Entech), Boeing, Acurex and Spectrolab.

In Europe, research institutes such as Technical University of Madrid, Catholic University of Louvain and IOFFE - Physical-Technical Institute of Saint Petersburg started research activities in the field of CPV in this period.

After a stagnation period that lasted from mid 1980s to the end of 1990s, this technology gained new interest owing to the highly efficient multi-junction devices based on gallium selenide that allowed achieving record efficiency of 43.5% in a short period, obtained in 2011 (triple cell by Solar Junction, with 418 suns).

The key role was played by companies such as Amonix (USA), in the field of cells and systems, SunPower (USA) for cells, and, more recently, Emcore and SolarJunction (USA) and Azur Space (Germany), which made commercially available concentrated cells with an efficiency higher than 40%.

Crystalline silicon technology is, for now, left in the domain of low concentration.

In the field of concentrated PV cells, Europe has a remarkable position both with respect to research and manufacturing, owing to Fraunhofer Institute near Freiburg that holds the world record in single-junction III-V cells, Technical University of Madrid, Imperial College, Ioffe Institute.

Numerous companies are involved in the manufacturing process, e.g. Azur Space (Germany) making cells and Concentrix (Germany), Albengoa, Solar, IsoFoton e GuascoFoton (Spain), Obsolicon Solar (Sweden) making modules/systems.

Italy also has a strong presence in this field, both with its research conducted by ENEA, CNR, CESI, RSE, Universities of Florence, Ferrara, Parma and companies including Beghelli, Angelantoni-Archimede Solar Energy, Alitech, C-Power, Converter, Ecoware.

Concentrated PV cells are currently considered a promising technology in the medium and long term, an area attracting strong international attention as indicated in Implementation Plan (2010-2012) and Technology Roadmap (2010-2020) of Solar Europe Industrial Initiative.

9.1) CPV Technology

The technology is to build the solar cells into concentrating collectors, that use a lens to focus the sunlight onto the cells.

As a result, less semi-conducting materials are used for solar cells, decreasing material costs while collecting as much sunlight as possible. Efficiencies are in the range of 20 to 30%.

Basic principles of the CPV technology are shown in Figure 9.a. The collecting area F_o (lens or mirror) concentrates the sunlight onto a small area F_c (solar cell).

Geometrical concentration factor is given by the ratio:

$$X = F_o / F_c > 1$$

The cost-efficiency of application is guaranteed by the use of collecting surfaces with a unit price considerably lower than the respective cell price.

In essence, the cell size is reduced, with a favorable economic effect as a result of reduced use of the expensive semiconductors.

Even though the cell area is scaled with the concentration factor, modules to not follow the same reduction factor since their surface area (or, more generally, the collecting area) keeps the dimensions comparable to those in flat PV.

Possible reduction of the surface area in these systems should be related to higher efficiency of module, which, in perspective, may be 10-15% or 20% higher than in flat PV modules, leading to reduction of the surface area by, theoretically speaking, same percentage.

In its most general application, the technology requires that sunlight- collecting area be always oriented towards the Sun, necessitating the use of systems for searching and tracking (trackers). A distinction should be made among low, medium and high concentrations. With different levels of complexity, these three types of concentration have reached different levels of industrial maturity.

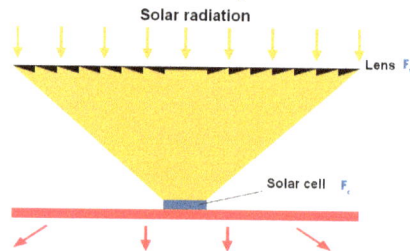

Figure 9.a – Operating principle of concentrated PV cells.

The first, generally based on c-Si cells, achieves concentration levels of the magnitude of several hundreds of Suns (with a concentration factor from 2 to 100X); the second and the third have concentration levels ranging from 100 and 300X and from 300X to above 1000X respectively; they use multi-junction (MJ) cells based on III-V materials and other technologies such as luminescent solar concentrators.

Multi-junction PV cells with high efficiency are made of multiple cells stacked on top of one another, each of them optimized for absorbing one portion of solar specter.

In the luminescent solar concentrators of the future (Fig. 9.b), the sunlight will be captured and directed in a concentrated form, owing to the use of thin film, applied in an appropriate way using luminescent nano-particles and appropriate structures to direct light.

Reduced in area and optimized for concentrator specter, the cells are arranged along the concentrator walls, i.e. the exit zone of the concentrated light.

Optical and tracking systems have various levels of complexity depending on the concentration type (Fig. 9.c).

In the case of low concentration, single-axis tracking is carried out (modules keep the inclination fixed, just following the Sun's azimuth) and optical systems generally employ simple flat mirrors.

In the case of medium-high concentration, dual-axes tracking is employed (directing toward the Sun according to azimuth and height); it employs an optical refractive system, such Fresnel lens or parabolic mirrors (in solar dish systems and linear concentrators) supported by secondary optical system that combines refraction and reflection to maximize concentrated light.

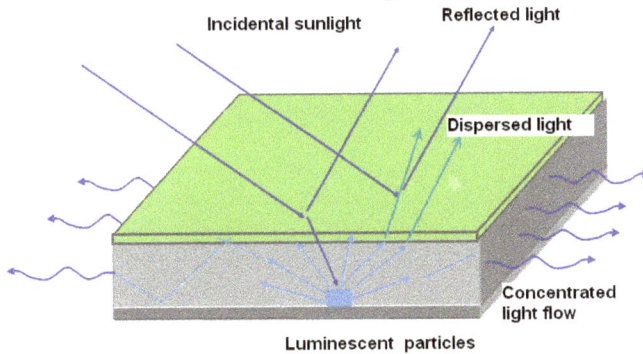

Figure 9.b – Diagram of a luminescent solar concentrator.

Figure 9.c - CPV system with single-axis (i). Dual axes with point focus modules (ii). Linear concentrator (iii)

It should be emphasized that optical concentrating systems operate solely based on the direct normal irradiation (DNI) components.

This implies loss of a portion of solar radiation, varying from 5% to 35% and more, depending on weather conditions (humidity, cloudiness, air pollution etc.) and latitude (in the equatorial belt and desert regions the DNI component is the most prominent). Regardless of losses, the high efficiency of cells allows module efficiency to exceed 30%, whereas system efficiency is over 25%, with excellent prospects to achieve conversion efficiencies over 34-35%.

An important characteristic of CPV is its predisposition to be used for cogeneration (PV-T).

Figure 9.d indicates an industrial prototype of a *solar dish* (ZenithSolar, Israel), part of a experimental device generating 70 kW$_p$ of output power and 140 kW$_{th}$ of thermal energy, installed in Yavne, Israel.

In certain conditions, such systems have recorded total energy (electric + thermal) efficiency up to 63%.

Figure 9.d – Concentrated system with a solar dish for cogeneration PV-T

The characteristics of concentrated PV systems, especially its weight and dimensions of its tracking system, have so far restricted this technology to applications in centralized plants of medium-large scale, necessitating mounting on the ground. Nevertheless, the cogeneration possibility makes it suitable for applications in places where the heat is used for industrial purposes or for desalination and solar cooling process.

The full affirmation of CPV should necessarily occur through structure and equipment simplification, which is the subject of ongoing research with a goal to finally provide at the same time:

o high efficiency,
o cost-efficiency of the investment,
o reliability,
o high level of product recyclability,
o using modest quantities of semiconductors, metals and other materials required to produce cells, modules and systems that are not different from the flat PV systems.

Therefore, concentrated PV systems may become a valid technological option for non-centralized applications, including architectural integration (BIPV).

To maximize the electricity generation, CPV modules need to be permanently oriented towards the sun, using a single- or double-axis sun-tracking system. Multijunction solar cells, along with sun-tracking systems, result in expensive CPV modules in comparison with conventional PV.

On the other hand, their higher efficiency and the smaller surface area of active material, as required, may eventually compensate for the current higher costs, depending on the evolution of costs and efficiency.

Because CPV modules rely on direct sunlight, they need to be used in regions with clear skies and high direct solar irradiation to maximize performance.

Picture 9.e - CPV system

In fact, third-generation technologies are yet to be commercialized at any scale. Concentrating PV has the potential to have the highest efficiency of any PV module, although it is not clear at what cost premium.

Other organic or hybrid organic/conventional (DSSC) PV technologies are at the R&D stage.

They offer low efficiency, but also low cost and weight, and free-form shaping. Therefore, they could fill niche markets (e.g. mobile applications) where these features are strongly required.

Technology	1st Generation		2nd Generation			3rd Generation		
	Single crystalline silicon (sc-Si)	Polycrystalline silicon (pc-Si)	Amorphous silicon (a-Si)	Copper Indium Gallium Diselenide (CIS/CIGS)	Cadmium Telluride solar cells (CDTe)	III-IV compound Multijunction, Concentrated PV (CPV)	Dye-sensitized (DSSC)	Organic or Polymer (OPV)
Best research solar cell efficency at AM1.5 [%]	24.7		10.4 Single junction 13.2 Tandem	20.3	16.5	43.5	11.1	11.1
Confirmed solar cell efficiency at AM1.5[%]	20-24	14-18	6-8	10-12	8-10	36-41	8.8	8.3
Confirmed PV Module efficiency at AM 1.5 [%]	15-19	13-15	5-8	7-11	8-11	25-30	1-5	1
Confirmed maximum PV Module efficiency at AM 1.5 [%]	23	16	7.1/10.0	12.1	11.2	25	-	-
Current PV module cost [USD/W]	<1.4	<1.4	~0.8	~0.9	~0.9	-	-	-
Market share in 2009 [%]	83	3	1	13		-	-	-
Market share in 2010 [%]	87	2	2	9		-	-	-
Maximum PV module output power [W]		320	300	120	120	120	-	-

Table 9.f - An overview and comparison of major PV technologies

10. Solar thermal energy

10.1) Existing technologies for low and medium temperatures

Basic technologies applied in solar thermal systems are glazed flat-plate selective and nonselective (FPC, *Flat Plate Collector*) collectors and vacuum tube collectors (ETC, *Evacuated Tube Collector*).

Besides them, there are also *uncovered collectors*, which mostly correspond to seasonal application.

10.2) Glazed flat-plate collectors

Glazed flat-plate collectors (Fig. 10.a) are very present and adaptable thanks to optimal annual energy efficiency and wide range of products in the market.

The principle of operation is the same as in a greenhouse: when sun's rays come into contact with a glazed area, a small number of them reflect back, whereas a larger number passes through the glass being absorbed by the collector.

Collector is heated and it releases energy in the form of infrared radiation (heat). With regard to heat, the glass acts as being opaque and it keeps energy (greenhouse effect). Frequently, there is a tendency that the glass reflects as low solar radiation as possible, and to be as opaque as possible for the heat released by the absorption material: that is, it is tried that all (or almost all) sun's rays enter the solar panel, and that the heat is then captured in the panel in order to be better delivered to the circulation fluid.

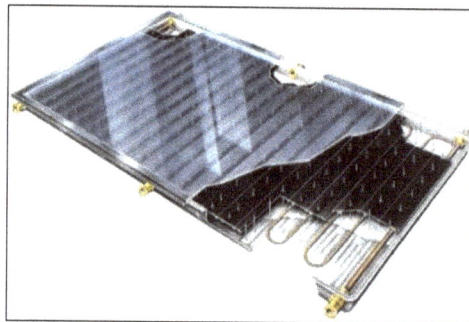

Figure 10.a – Example of a glazed solar collector

In order to increase the efficiency of heating induced from solar energy, absorption plate has a special surface coating so as to be transparent for the sun's rays and at the same time opaque for the infrared radiation emitted in the metal plate.

By this technique, as well as by glazing, high values of solar radiation absorption are reached, along with low release of infrared energy (*selectivity*).

According to the type of construction, there are numerous solutions differing in selectivity of absorption, material used (copper, stainless steel and anodized aluminum) and appropriate use in the facilities with forced or natural circulation (these latter are cheaper, more reliable, but at the same time more difficult to be installed in architectural terms, considering that it is necessary to place a storage tank above the panel, but in its close proximity).

Although there are various variants in the market, usual collector dimensions require an area of about two square meters, with the longer side typically extended on the length of two meters.

10.3) Structure of glazed flat-plate collector

In order to better understand the physical principles on which solar collector operates, it is necessary to give consideration to glazed flat-plate collectors, since these are the simplest and the most present in the market.

These collectors include several elements (Fig. 10.b):

- *Solar heat absorber:* It consists of a metal plate with darkened face directed towards solar radiation; in its interior there are pipes through which primary circuit fluid flows, which is to be heated by the sun.

- *One or more glass plates* or plates made of some other transparent material: They are located above the absorber protecting the device and allowing the passage of incident sun's rays; on the other hand, during heating, the absorber releases energy in the form of infrared radiation, but with regard to this energy the glass behaves as it is opaque and, hence, it keeps that energy inside (greenhouse effect).

- *Thermal insulator:* It is located in the bottom or lateral part of the panel, and its purpose is to reduce dispersion of the heat for heat conduction.

- *Design:* It is made of sheet metal or plastics, closing the panel and protecting it from weather effects.

All the elements are connected by a metal structure (typically aluminum structure), which helps assembling the parts and provides for hardness and stability of the panel.

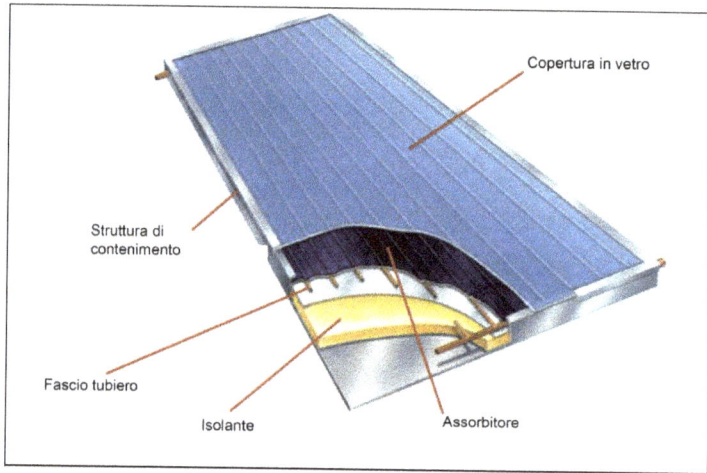

Figure 10.b - The structure of a flat-plate solar collector

Absorber is a basic component of the solar collector and its properties are to a great extent determined by the total efficiency.

It is characterized by a special design technology, as well as the ability to absorb the solar radiation. In that respect, the value of the absorption coefficient α is important in the wavelength range of the solar spectrum, which is equal to the absorption coefficient α, i.e. the ratio of the absorbed and incident radiation on the surface for a given wavelength (total value between 0 and 1).

At the same time, absorption coefficient is the function of the angle at which solar radiation falls on the absorber surface. On the other hand, absorber is characterized by a certain value of the emission coefficient ε, which is defined as a ratio of the radiation emitted from the surface and the radiation emitted by a black body, at the same temperature, for a given wavelength, and hence its order of magnitude is also from 0 to 1.

In case the coefficient value increases in the range of wavelengths specific for the thermal radiation emitted by a plate, depending on the temperature, considerable losses may occur due to infrared radiation.

With the aim to reduce those losses, a selective finish is applied with a high value of solar spectrum absorption (0,3 μm < λ < 3 μm) and low emission values in the wavelength range, with the temperature of the absorption plate (3 μm < λ < 10 μm).

With regard to selective surfaces, the basic requirement is their hardness or the need to maintain surface properties in the time period corresponding to the average life cycle of a collector (ranging from about 15 to 20 years).

In respect of the method of taking away the heat absorbed by the plate, there are various designs that can be reduced to three types:

- Water tube plate.
- Roll-Bond plate.
- Plate with parallel metal sheets.

Water tube plate – In this type of absorber, the tubes are arranged differently than in a flat plate and can be parallel or spiral (See Figure 5.10).

Fundamental condition is to achieve maximum contact between the tubes and plate in order to allow heat-conveying fluid to absorb a larger fraction of the heat accumulated in the plate.

The efficiency of the surface that accumulates heat varies depending on the plate thickness, number of tubes and the type of metal; typically, copper or aluminum is used for the plate and copper for the tubes.

The larger the number of tubes per m², i.e. the smaller the distance between them, the larger is the amount of accumulated heat.

In 1959 Bliss discovered that as a trade-off between the price and efficiency, optimal distance between the tubes amounted to 10-15 cm for a plate with a thickness of 0,25 mm.

Optimal tube diameter is obtained through a trade-off between the tube price, which decreases with the reduction of diameter, and load losses, which decrease with the increase of diameter.

Finally, it should be also said that absorption surface consists of finned tubes with fins having the function to store and take away the heat towards the tubes.

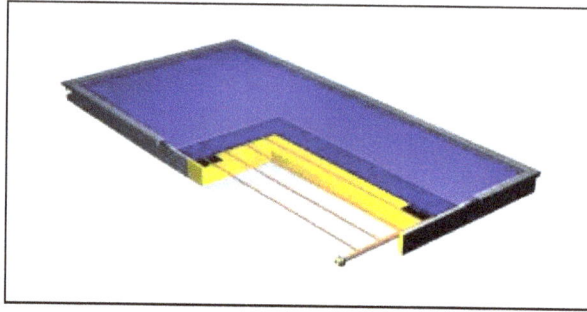

Figure 10.c – Example of a glazed flat-plate collector with a water tube plate (Source: www.energie-rinnovabili.net)

Roll-Bond plate – The plate is named after the process of manufacturing invented in USA and consisting of hot rolling welding of the two metal plates. On one plate, a path for heat-conveying fluid is impressed by serigraphy.

Since according to such plan, welding is not achieved between the two plates, it is possible to inject pressurized liquid or gas, which causes formation of a channel with an oval cross-section.

The effect of thermal exchange between the plates and fluid is higher than in other systems, owing to complex circuits that are adequate for various needs; the channels obtained in such a manner are an integral part of a solar exchanger, and thus the problem of welding is eliminated.

However, some issues such as, for instance, the internal overpressure, which causes separation of metal sheets and outflow of heat-conveying fluid, have limited the application of this technique.

Plate with parallel metal sheets – Absorption plate consists of two opposite metal sheets at the distance of several millimeters, between which the heat-conveying fluid flows.

The advantages of this technology include the possibility to apply stainless steel, which besides an excellent corrosion resistance also allows the application of very small thicknesses, at the same time not compromising the hardness.

The special representative of this technology is the absorber specialized at the University in Lausanne characterized by the chess board geometry.

This absorber is constructed by hard welding of two metal sheets, on which previously the squares have been impressed, alternating with each other at the distance equal to the half of their length, in the both directions; in this way, 97% of the surface exposed to solar radiation will be in contact with water.

Geometry created in such a manner ensures almost uniform distribution of the water flow inside the absorber, along with an excellent thermal exchange coefficient.

10.4) Covering

As it has already been said, if the absorbing surface of a collector is not protected, it cannot reach the temperature required for an efficient operation since, as it is slowly heated, at the same time it loses the heat.

To avoid these solar losses, the collectors are covered with panels made of glass or transparent plastics.

One portion of the solar energy that reaches the transparent panel is reflected, one portion is absorbed, whereas the remaining portion is transferred to the absorption plate.

Optical transmittance or transmittance coefficient τ depends on the composition, surface condition and thickness of the transparent panel, as well as on the incident angle and radiation.

For the purpose of providing maximum solar energy reception, it is necessary that τ has the highest possible value for the entire solar spectrum. Since the absorber then radiates again at increased wavelengths ($\lambda > 3$ μm), the transparent panel has to pass through that type of radiation, so that the greenhouse effect is ensured.

10.5) Vacuum tube collectors

In case their areas are equal, vacuum tube collectors allow for better average seasonal efficiency as compared to glazed flat-plate collectors, due to annulling the thermal losses in the external environment owing to the gap that is maintained under vacuum and absolute pressure with the order of magnitude of 5×10^{-3} Pa.

In connection with Figure 10.d, vacuum tube solar collector typically consists of two glass tubes (*all-glass* type of collector), closed at the opposite ends, whereas the space between them is evacuated.

The exterior surface of the inner tube has a selective coating (usually aluminum nitride Al-N/Al), characterized by a high value of solar spectrum absorption and low value of infrared emission.

Selective layer can be obtained by galvanic technique or in the vacuum. In the glass tube, there is a copper tube in the form of the letter U, with aluminum fins adhering to the inner surface of the glass tube, thus ensuring thermal contact with the absorber and transfer of heat in the copper tube where the conveying fluid flows.

Keeping in mind its cylindrical shape, in some cases, collector has parabolic reflectors of CPC (*Compound Parabolic Concentrator*) type, which allow solar radiation to concentrate at the rear side of the tube and not to be directly intercepted by the vacuum tube, since in that case it would be lost.

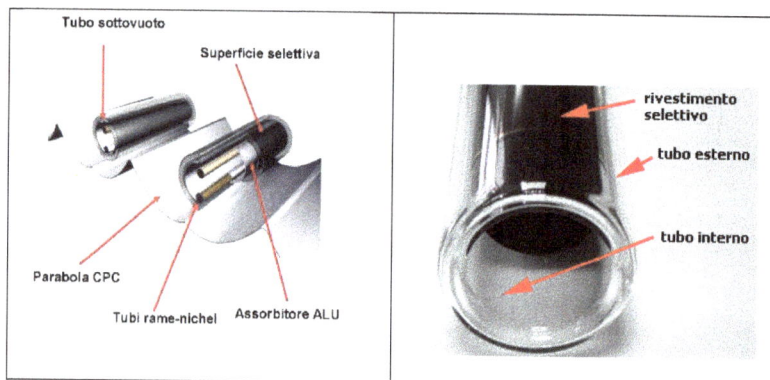

Figure 5.d – Vacuum tube solar collector, all glass type

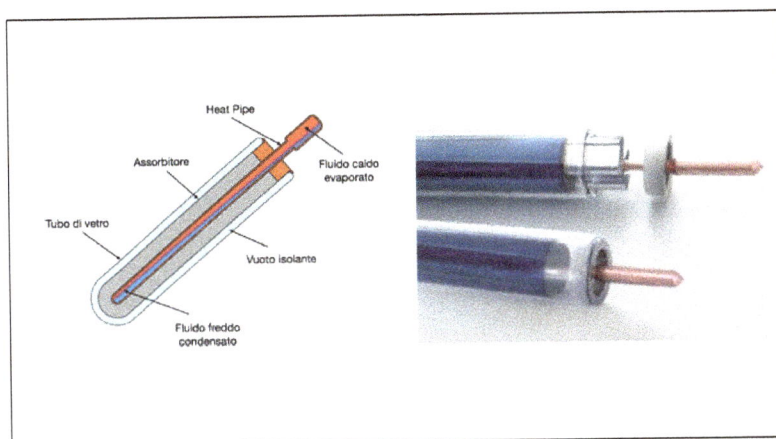

Figure 10.e - Vacuum tube solar collector, heat-pipe type

Instead of the tube in the shape of the letter U, many vacuum tube collectors use "heat pipes" consisting of the pipes made from a high thermal conductivity material (usually copper or aluminum), sealed at the ends and containing low-boiling fluid, which can be ordinary distilled water closed at a pressure lower than atmospheric pressure.

Depending on the fluid type, quantity, and absolute pressure, various range of the operating temperature of heat pipe is obtained.

Its operation is based on the constant evaporation and condensation of low-boiling fluid, which in the lower section of the heat pipe is in contact with the heat source consisting of a cylindrical absorber exposed to solar radiation, while in the upper section (*superior bulb*) it comes into contact with the heat-conveying fluid flowing through the bulbs and causing condensation and delivery of heat.

Upon condensation, due to the gravitation, the low-boiling fluid goes down the interior walls of the heat pipe to lower parts where a new cycle starts.

Although they have the advantage of better maintenance of vacuum, vacuum tube collectors of *all-glass* type have less optical efficiency because of the double glass; in addition, there is a possibility that humidity and dirt reach the absorption tube and cause corrosion.

These problems can be partially solved by using another type of vacuum tube (known as *metal in glass*), which consists of a glass enclosed in a finned copper tube, of *heat-pipe* type, with the use of glass-metal welding.

In this configuration, corrosion problem is eliminated and at the same it offers higher optical efficiency since only one glass is used. On the other hand, the structure is fragile and long-term maintenance of vacuum is less reliable.

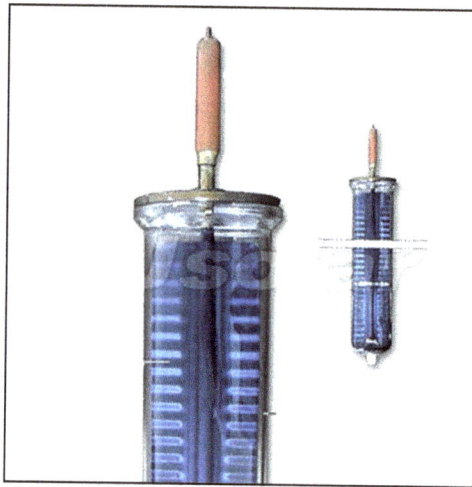

Figure 10.f – Detail of a heat-pipe of "metal in glass" type

Vacuum tube collectors have a high stagnation temperature, i.e. the value of the average temperature of heat-conveying fluid at which the energy equal to losses is absorbed (efficiency equals zero).

It depends on radiation and outdoor temperature; usually, when speaking of stagnation without specifying these values, it is thought of solar radiation value of 1000 W/m² and external temperature of 30°C.

The higher the stagnation temperature is, the higher is the efficiency at high temperatures.

For vacuum tubes, stagnation temperatures are about 220-250 °C ; efficiency is about 60% for the operating temperature of 100 °C and about 40% for the operating temperature of 150 °C.

This type of collectors were previously used at higher temperatures than those that could be reached with flat-plate collectors, while today they are present in a wide market, and with the introduction of double tube collectors, vacuum technology becomes rather interesting in Italy.

10.6) Unglazed collectors (or uncovered collectors)

Unglazed collectors (or uncovered collectors) have low price and are primarily intended for the summer use.

These are usually made of plastics (PVC, neoprene or polypropylene), but of metal as well (typically treated steel with a selective coating, see Fig. 10.g), where the absence of glass covering causes excessive losses so that they cannot be used in the winter period.

In the majority of cases, heating water flows directly through the collector and thus the costs and complications of installing an exchanger are avoided.

This is an ideal solution for public baths, covered swimming pools, camp and all residential buildings which have the need for sanitary hot water in the summer period (Fig. 10.g).

The second application, technically interesting, but not so widely spread is the application of open collectors as evaporators, with heat pumps assisted by helium, where this type of collectors is more energy efficient since operating temperatures are generally lower than outdoor temperature.

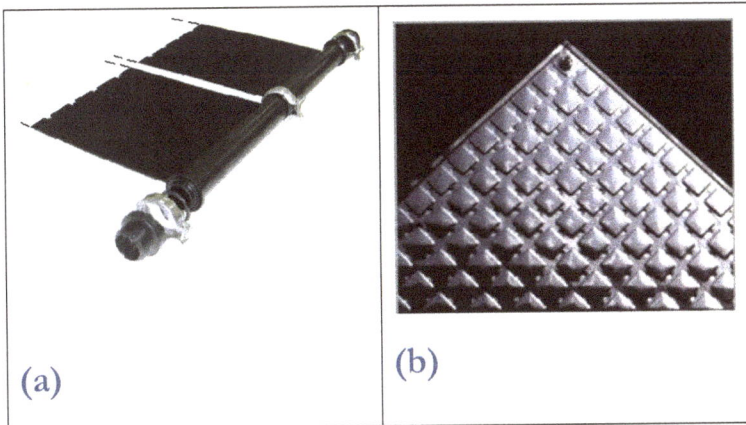

(a)

(b)

Figure 10.g – Example of open collectors made of plastics (a) and metal (b)

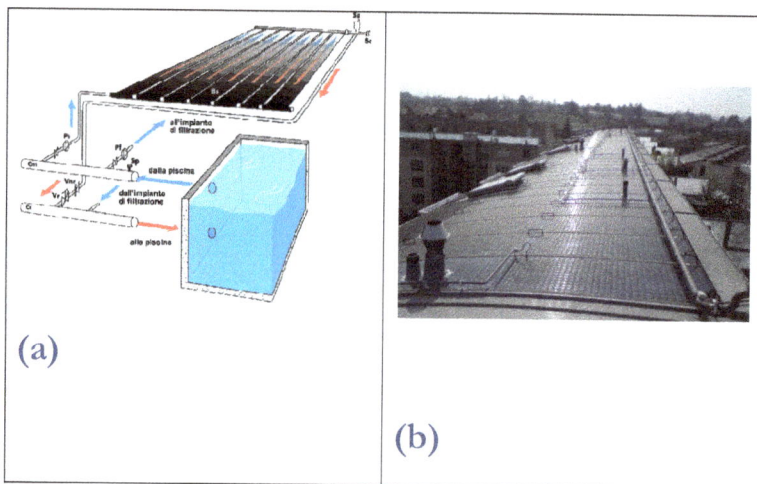

(a)

(b)

Figure 10.h – Examples of using uncovered collectors for heating water in a swimming pool (a) and sanitary hot water production (b)

10.7) Energy source of a collector and efficiency curve

As it has already been said, the function of a solar collector is to convert solar energy into heat energy, which is transferred to the fluid flowing in the pipes.

Figure 5.16 illustrates schematically the energy flow in a solar thermal collector and the basic elements of energy balance.

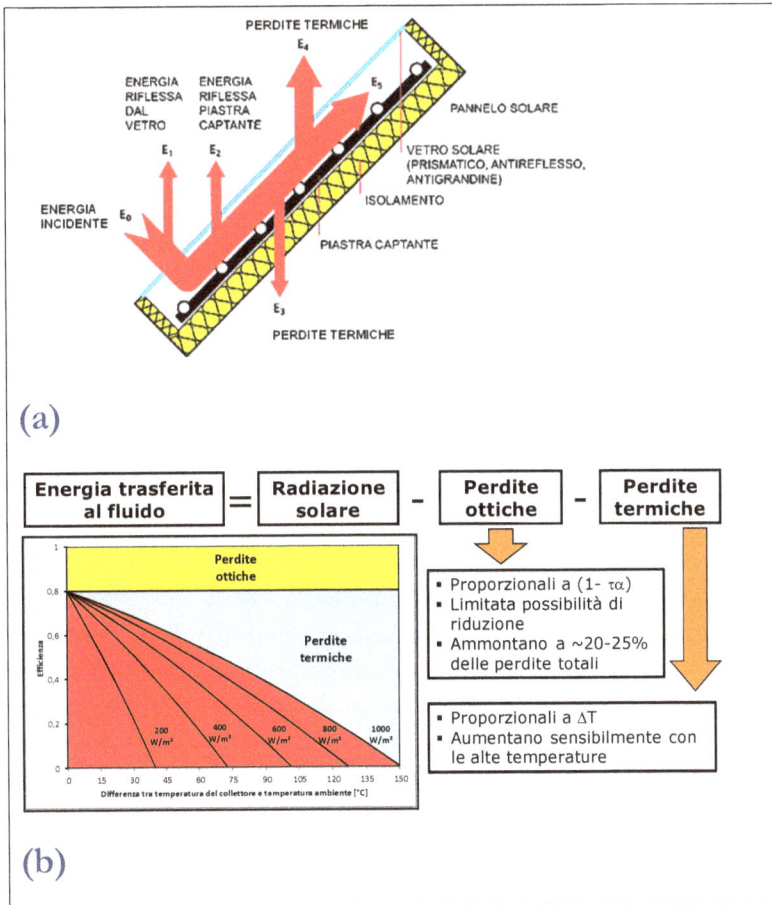

Figure 10.i – Schematic diagram of energy flow in a solar collector (a) and energy balance of a solar collector (b)

By the application of thermal balance on the collector, it is obtained:

$$GA_c \cdot (\tau\alpha) = Q_u + Q_p + Q_{acc}$$

Where:

- ✓ G total radiation on the collector plane (W/m²)
- ✓ A_c collector area (m²);
- ✓ $(\tau\alpha)$ effective product of transmittance and absorption.

It is not a simple product of two elements, since it is necessary to take into account multiple reflections between the plate and glass; therefore, this two elements are in the parentheses and there is no multiplication sign between them;

- ✓ Q_u useful thermal power transferred to heat-conveying fluid (W);
- ✓ Q_p collector power lost to the external environment (W);
- ✓ Q_{acc} heat energy accumulated in the collector per unit of time (W); if stationary mode is achieved, this element will disappear.

Current collector efficiency is a ratio of useful power transferred to heat-conveying fluid and global power acting on the transparent collector surface:

$$\eta = \frac{Q_u}{A_c G}$$

Without going into detail with regard to the mechanisms of conveying heat between the plates, fluid, glass and environment, it can be shown that it is possible to express radiated heat as the product of the collector area A_c and the coefficient U_c (W/m²K) and the difference between the average temperature between the plates and external temperature:

$$Q_p = A_c U_c (T_p - T_a)$$

The coefficient U_c takes into account heat losses from the absorbing plate due to convective exchange with air and exchange due to radiation towards the sky through the plate – glass interspace.

The average temperature of the plate is hard to measure and calculate. Hence, the average fluid temperature (T_m) is preferably used. It can be shown that the useful absorbed power of the collector amounts to:

$$Q_u = F' A_c \left[G\,(\tau\alpha) - U_c(T_m - T_a) \right]$$

where F' is the efficiency factor of the collector that represents the ratio between the effective power conveyed to fluid and maximum conveyable power that would be obtained in case that the average temperature of the plate equals the average fluid temperature. Efficiency factor F' increases with the increase of the thickness of absorbing plate and thermal conductivity.

In contrast to that, F' decreases with the increase of the loss coefficient U_c and with the increase of the distance between the pipes.

Collector efficiency that is defined as $\eta = \dfrac{Q_u}{A_c G}$, can be written as:

$$\eta = F' \cdot (\tau \alpha)_n - F' U_c \frac{(T_m - T_a)}{G}$$

In should be noted that the effective product $(\tau \alpha)$ that occurs when expressing the useful power depends on the incident angle of direct radiation.

However, on the efficiency curve shown above, the product $(\tau \alpha)_n$ appears, which has been determined in the conditions of the action of direct solar radiation on the collector plane at the right angle. The ratio between $(\tau\alpha)$ and $(\tau\alpha)_n$ is defined by the following formula:

$$K_{\tau\alpha} = \frac{(\tau \alpha)}{(\tau \alpha)_n} = 1 - b_0 \left(\frac{1}{\cos\theta} - 1 \right)$$

where θ is the incident angle, and b_0 is a specified parameter of the collector that is to be experimentally determined. $K_{\tau\alpha}$ is defined as "incident angle modifier".

Based on the previous equation, it follows that if the loss coefficient U_c is constant, the collector efficiency varies linearly with the parameter $(T_m - T_a)/G$.

In addition, we can see that the efficiency decreases with the increase of the input fluid temperature, since the heat losses are increased; it decreases with the increase of the temperature difference between the fluid and ambient, and increases with the increase of solar flux G.

Actually, U_c is not a constant value, but it depends on temperature. In this case, it can be shown that the following formula is applicable, which is normally used to express the variation of the efficiency of a thermal solar collector in the function of its operating temperature that is in accordance with meteorological-climatic conditions.

$$\eta_c = \eta_0 - a_1 T_m^* - a_2 G \cdot (T_m^*)^2$$

The symbols used have the following meaning:

η_0 a constant which gives the value of efficiency for T_m^* equals 0, which is also called optical efficiency;

G   total radiation on the collector surface (W/m²);

a_1, a_2 specific constants – non-dimensional – of the collector that take into account thermal losses to external environment.

Finally, with the symbol $T_m^* = (T_m - T_a)/G$ "reduced temperature" is designated, expressed in m²K/W, which allows representation of the efficiency curve independently on climatic conditions that have enabled its experimental determination.

The graph in Figure 5.17 shows the typical variation of the efficiency curve for different types of the above described collectors.

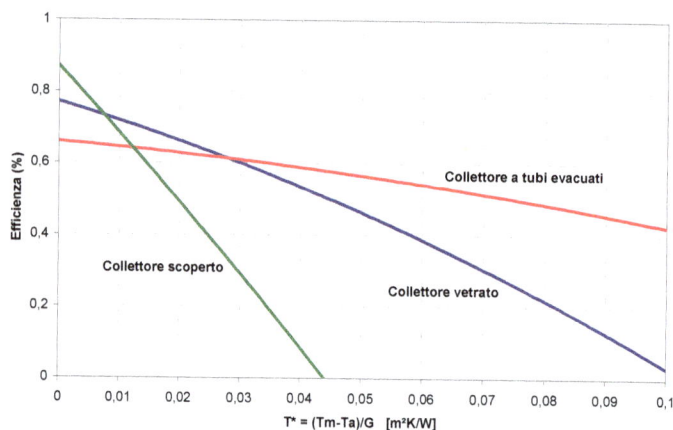

Figure 10.1 – Typical variation of the efficiency curve for different types of thermal solar collectors

11. Solar thermal energy applications

Thermal solar technologies find increasing application and can considerably contribute to the final consumption for heating.

In Europe, 49% of the final consumption is related to heat energy. Out of that, 34% is heat at low temperatures, and 61% of the total demand for heat at low temperatures is related to residential needs.

In Italy, although these percentages are lower, only the consumption of the final energy for household use amounts to about 31 Mtep of the total of 125.

Hence, the solar energy represents a technology that could significantly affect the achievement of the objectives of the national Action Plan in which covering in the value of 17% of final consumption is foreseen by 2020 with the energy obtained from renewable sources.

Based on the data published by the ESTIF (*European Solar Thermal Industry Federation*) in its last report on the condition in the European market of solar heat energy, today, in the entire Europe, total more than 37 million m² of solar thermal collectors are installed (corresponding to the values of the installed power of about 26,3 GW$_{th}$), and out of that, 35% in Germany.

Due to economic crisis, the European market of solar heat energy recorded a drop in 2011, although less than in 2010, and it ranges about the values of approximately 3,6 million m² or 2,6 GW$_{th}$.

Italy is the second market in Europe, where about 425.000 m² (about 290.000 kW$_{th}$) are installed annually.

However, it has recorded a drop of about 13% as compared to 2010, and the total installed value amounts to 3,1 million m², which is far away from the goals of the National Action Plan.

Long expected increase concerning the facilities for the industrial application has not led to major changes so far comparing to the previous years, although this market segment is promising with regard to the application of medium temperature solar devices.

Despite the economic crisis, there are favorable conditions for the development of production units for solar panels due to the rise of the prices of fossil sources and ecological issues, which impose an increasing use of less polluting sources and the least possible effect on the climate.

Technological progress is also very important considering that solar heating technology nowadays is very competitive with its numerous applications, particularly in the household sector, both for hot water generation and for room air conditioning (winter and summer time) with the use of combined systems and innovative solar-cooling systems that, among other things, can utilize low temperature solar thermal systems and concentrating collectors.

Sanitary hot water production is the most frequent application of low temperature solar thermal systems.

In the solutions for summer use, such as for instance public baths and associated structures, application of uncovered collectors is possible.

For partial or complete load covering in case of insufficient solar radiation, storage and integration systems are required.

Glazed flat-plate collectors and vacuum tube collectors can be also used for heating, particularly within low temperature systems.

Solar air-conditioning system (autonomous and/or assisted) is one of the promising applications of solar thermal systems, which allows significant savings in primary energy. Production systems for cooling services are adapted for the use of solar energy owing to correlation between the existing radiation and summer demand for air conditioning.

Cooling and air conditioning are mature sectors, but still offer wide innovation potential. Solar-cooling solutions represent a sector of vast development potential that is to reach its economic competitiveness and technical maturity.

In this area, the activities of research and development are directed to improving control systems, thermal storage and heat-conducting means, as well as to obtaining maximally efficient and compact units.

The development of small helium air-conditioning devices (< than 10 kW_{th}) can become a solution with the greatest energy and ecology benefits that would meet ever growing demand for small electric decentralized air-conditioning devices and at the same time cover the needs both for heating and air-conditioning.

The second area of application is solar desalination. In various countries around the world, new system is tested for water desalination and treatment with the use of solar thermal systems for the purpose of developing small capacity plants, thus overcoming technical incompatibility with the present desalination systems.

One more field of application, with a strong development potential, is heat energy production for industrial branches, with a constant demand for low and medium temperature heat (up to 250 °C), as well as technical possibility of including a solar system into an existing industrial process.

As it can be seen in Table 11.a, a considerable fraction of heat is required for industrial processes at temperatures lower than 200 °C, which can support the integration of solar systems with flat-plate collectors or vacuum tube collectors for lower temperatures and linear parabolic collectors for higher temperatures.

Solar thermal systems are the most present in the food sector, i.e. production of *food & beverage*, textile sector, chemical sector, as well as in desalination plants of water for household and industrial use.

The greatest potential lies in the agricultural/food sector where the heat energy is used for sterilization, fermentation, pasteurization and cooking of food.

Starting from the estimate of the national agency ENEA for the European market, there is an intention to cover the needs for heat energy with the use of solar technological solutions in the Italian industry in the value of 3÷4% by 2020, which would correspond to the installation of about 4,8 million m² of solar surfaces with power above 3,4 GW_{th}.

Table 11.a – Specific temperatures of some industrial processes that can be implemented with medium temperature solar technology solutions.

Industrial branch	Production process	Temperature (°C)
Food and beverage	Washing	80- 150
	pasteurization	80-110
	sterilization	130-150
	drying	130-240
	cooking	80-100
Industrial production of plastics	extrusion and drying	150-180
Chemical industry	thermal treatment	150-180
	boiling	95-100
	distillation	110-300
	drying	150-180
Paper industry	bleaching and drying	130-180
Textile industry	Washing	80-100
	thermal treatment	80-130
	bleaching	60-100
	dying	100-160
Industrial laundries	steam cleaning	150

Despite the fact that energy demand in the industrialized countries represents about 30% of the total national demand and that two thirds of that is the demand for heat energy, mostly at the temperatures lower than 250 °C, today the use of solar systems for producing the heat for industrial purposes is modest and experimental by nature.

Presently, the market of medium temperature solar thermal energy is still in the initial phase of its development, half the way between the sector of low temperature solar thermal energy (technology that can be considered mature and widely spread) and solar technologies for the production of power (thermodynamic solar systems), with market struggling for survival despite a strong development potential and an active presence in the Italian industry.

The development of concentrating technology and small size systems (mini and micro CSP, systems based on the linear Fresnel concentrators) would surely open the path for a wide application and commercial presence.

That is, these technological solutions, due to their smaller dimensions as compared to large size thermodynamic systems, are very attractive for commercial and industrial consumers.

In addition, these systems are particularly suitable (and economically competitive) for co-generation or three-generation applications, where the heat that has not been used in the electricity generation can be used for supplying the processes with medium temperature and/or for cooling with absorption cycles.

Considering the small dimensions of these systems, electricity generation would be transferred to high-efficient micro-turbines and organic fluid turbine (ORC - *Organic Rankine Cycle*), which are particularly suitable to utilize medium temperature heat produced in solar concentration plants.

11.1) Sanitary hot water production

For the thermal use of solar energy in low temperature applications, the following is required (Fig. 11.b):

- *Capture system* consisting of a solar collector that absorbs solar radiation and convert it directly into heat energy conveyed through an appropriate heat-conveying fluid ;
- *Storage system* consisting of a fluid tank, and having the basic function to reduce the variability of energy and to deal with non-conformities between the available energy and energy demand due to weather conditions;

- *Hydraulic circuit* connecting the collectors and storage with appropriate control system of fluid circulation; considering that heat-conveying fluid is almost always different from the one used by the consumer, it is necessary to foresee an exchanger of heat between the two fluids;

- *Integration system* in order to satisfy the periods of less solar heating.

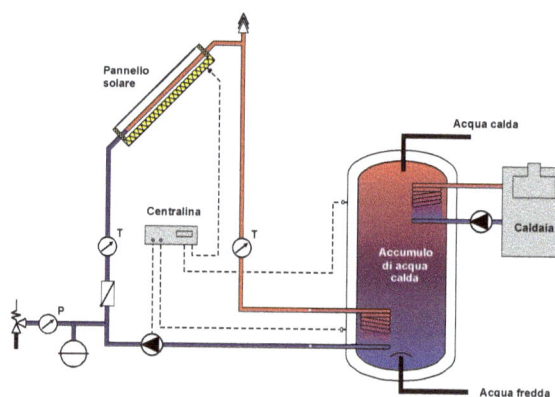

Figure 11.b – Sanitary hot water system

11.2) Classification of the systems

In the market, there are various types of solar systems, which take into account various consumers' needs.

The first major classification of solar systems is as follows:

❖ *Natural circulation systems (factory-made):* in these systems, also known as *factory-made* circulation systems for heat-conveying fluid, physical law is used according to which water rises when heated since its density is reduced and thus a natural convective motion is established; in order to contribute to that natural circulation, the tank has to be placed above the collector so that fluid is moved due to the temperature difference between two branches of the circuit. An example of a natural circulation system is shown in Figure 5.19a.

In this and in the following figures, auxiliary source is an electrical source previously used for small-sized devices intended for one household.

Since the use of electricity represents wasting in terms of energy, today the majority of plants are *factory-made* connected in series with a natural gas fired boiler. In larger plants, the boiler supplies a winding (spiral) located in the upper section or a solar water heater (for devices with one tank) or an auxiliary boiler for devices with double tank.

Special types of *factory-made* systems are integrated collector-storage systems (ICS, *Integrated Collector Storage*) where the section of solar capture (collector) and the section of thermal storage (tank) make a united device (See Figure 5.19b).

In this case, solar radiation directly heats the water in the collector and consumer can further use it.

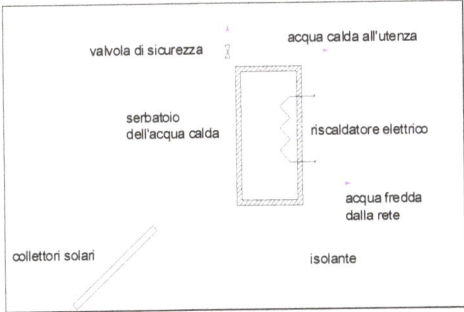

Figure 11.c – Natural circulation system

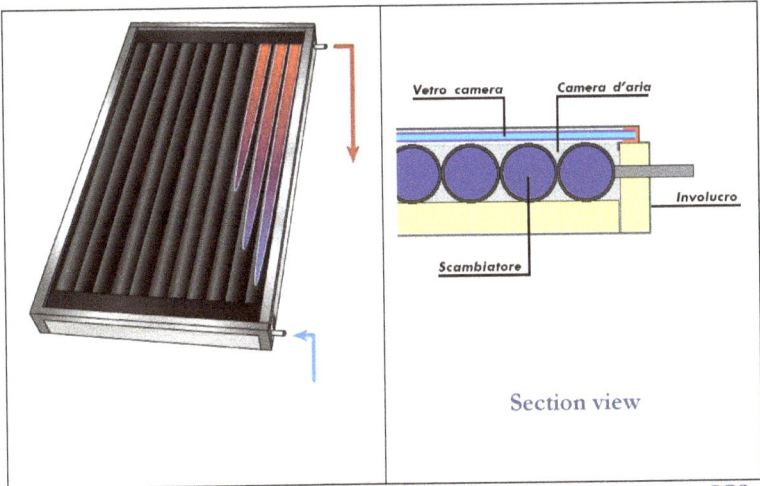

Figure 11.d – Example of an integrated system collector-storage, ICS

❖ *Forced circulation systems (Custom-built)*: These are used when it is not possible to place a tank above the collector and for collector areas larger than 4 m² since in that case natural circulation is not adequate to ensure the required supply of heat-conveying fluid; in this case, differential thermostat is used to control the temperature difference between the fluid at the collector output and water in the tank: when this difference exceeds a minimum limit, circulation pump is activated and the heat of heat-conveying fluid is transferred in the tank.

An example is shown in Fig. 11.e.

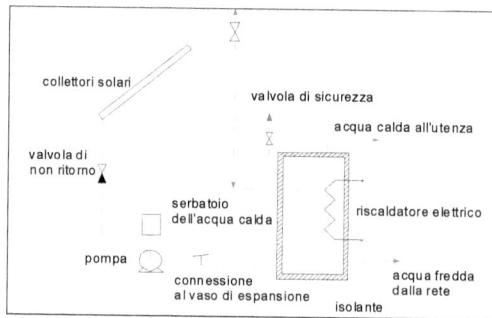

Figure 11.e – Forced circulation system

Solar systems can be classified according to the method of conveying heat to storage tank:

• *Open circuit systems:* These are the systems where heat-conveying fluid, which flows through collectors, is water intended for consumption; these systems are not used in Italy since, due to climatic conditions, the use of antifreeze solution cannot be avoided at low temperatures in the winter period or during the night. These systems are not equipped with a heat exchanger, so that they have higher efficiency since the fluid circulating in the panels can be used at the maximum available temperature and at a lower cost (Figure 11.f).

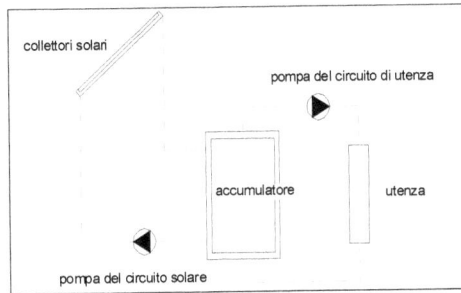

Figure 11.f – Open circuit system

- *Closed circuit systems* are systems predominantly used in Italy, where heat exchanger allows conveying the heat from heat-conveying fluid at the collector output (usually water mixed with antifreeze solution) to water in the tank (Fig. 11.g). Heat exchanger can be placed inside the tank or on its exterior.

Various studies have shown that the use of an external exchanger improves annual device performance; however, in the solar system selection, attention should be paid to the fact that an external exchanger has a higher price and size. Several types of heat exchangers are used; in the forced circulation systems, winding channel exchanger is usually used, located in the bottom part of the tank. In natural circulation systems, coated exchangers are used; heat-conveying fluid enters the tank and "splashes" the external area of an identically shaped vessel, which contains water for consumption; thus the heat exchange is more efficient.

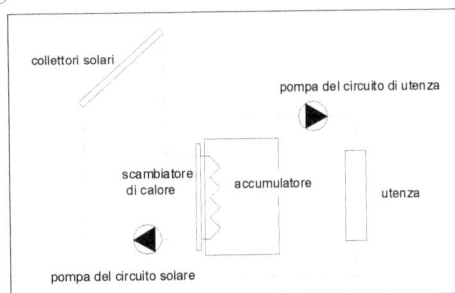

Figure 11.g – Closed circuit system

Further classification of the solar systems can be as follows:

- *Solar-only systems:* These are the systems where water is heated only with the use of a solar source; they are used in the areas where there is considerable solar energy or only in the summer period;

- *Integrated energy systems:* These are the systems used during the whole year and also in the countries where, due to climatic variations, the full energy demand cannot be met with only a solar source. There are various solutions depending on application. There are systems with one tank and systems with two tanks; the first solution is applied for *factory-made* systems, which in some cases have electric resistance within the tank that acts in case of insufficient contribution of heat energy from the solar source. The second solution with two tanks offers a better efficiency, but higher costs too. The first tank is heated by sun's rays and brings water for use into the tank, which is heated by a conventional boiler. In that case, the operating temperature of the fluid in collectors has lower values, and therewith the capture is more efficient. Solar systems with two tanks are medium - or large-sized and can satisfy the needs of several households.

11.3) Hydraulic circuit

The purpose of the hydraulic circuit is to transfer thermal energy collected in the collectors to consumers.

Its basic components are:
- Device tubes
- Circulation pump, it is a forced circulation device
- Storage tanks

Before describing these components, it is necessary to address the methods of connecting the collectors.

Collectors are usually interconnected in parallel, by connecting the outputs and inputs; usually such an assembly consists of 2 do 6 collectors; generally, the number of 6 collectors is not exceeded, since the experiences have shown that otherwise the distribution of flow among different collectors is not uniform, with some collectors supplied from below, and hence having lower efficiency.

The assemblies are interconnected in parallel. In order to evenly distribute the carrying capacity among the assemblies, a reverse return circuit can be used or return from the last to the first assembly can be provided, taking care to include balancing valves for each assembly, since in this case the circuit is not self-balanced. Figure 11.h illustrates the both types of connection.

Centrifugal pump allows the circulation of heat-conveying fluid between the collector assembly and storage water heater.

In order to adopt a methodology of appropriate dimensions, care should be taken of the following:

- Reverse return circuit is self-balanced, so that driving power of the pump (usually centrifugal, with several speeds) is obtained by the calculation of load losses of any closed loop within the circuit that includes the pump.

- In the circuits with balancing valves, prevailing power is obtained by the calculation of load losses for the most unfavorable loop, and that is the one which includes all the assemblies.

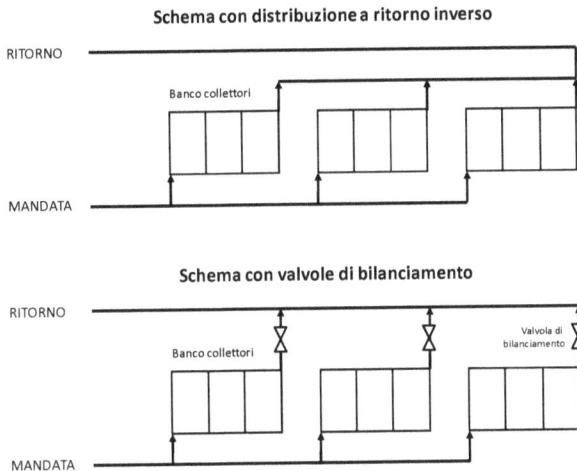

Schema con distribuzione a ritorno inverso

RITORNO

Banco collettori

MANDATA

Schema con valvole di bilanciamento

RITORNO

Banco collettori

Valvola di bilanciamento

MANDATA

Figure 11.h – Distribution with reverse return and balancing valves

Heat energy collected by collectors is stored in storage tanks. The tanks for sanitary hot water and drinking water have internal corrosion-protection finish and maximum operating pressure of 6 bar.

Each individual tank is equipped with:

- Automatic air vent
- Expansion tank with interchangeable supply-type membrane
- Safety valve and thermal take-away
- Temperature indicator for sanitary hot water
- Air gauge for indicating the system pressure
- Pressure reducer if required

11.4) Control system

In the forced circulation systems, a pump is required only when it is possible to convey thermal energy from the collector to storage.

Namely, in cases of low insulation and high temperature of water in water heaters, it may occur that the temperature of heat-conveying fluid at the collector output is lower than the stored temperature and then the water heater will cool, dissipating heat energy into the environment.

The control system consists of a central control unit for the pump and two temperature sensors, one at the collector output, and the other in the bottom part of the water heater.

When the temperature difference between the collector output and water heater exceeds a certain level (usually from 5 to 10 °C), the central control unit activates the pump; if that difference drops below the minimum value (2-3 °C), the central control unit deactivates the pump.

Temperature jumps are selected carefully since a low value may cause oscillations with constant switching the pump on and off, which will overload the pump; high values cause the reduction of collected energy.

11.5) *Criteria for determining solar plants dimensions*

The criteria for determining the size of the systems for sanitary hot water production, which represent the most widely spread application of low temperature solar thermal energy, are given below.

In manufacturing of sanitary hot water systems, above all, it is necessary to know the end user, i.e. if it is a household, hospital, hotel, army barracks, etc.

Table 11.i shows required liters/days for the given application in the most frequent cases.

Based on the amounts, operating temperature and temperature of water in the system (network), thermal load in MJ/day can be determined – which is shown in Table – and, therefore, monthly load that will be used below in the technical-economic estimate.

Table 11.i – Loads for sanitary hot water production

	lt/day per person	MJ/day per person	kWh/day per person	NOTE
Houses	50	6,9	1,92	-
Hospitals	60	8,29	2,30	per bed
Nursing homes	40	5,52	1,53	-
School	5	0,69	0,192	-
Military	30	4,14	1,15	-
Industries	20	2,76	0,767	-
Offices	5	0,69	0,192	-
Camping	30	4,14	1,15	per person
Hotel high class	160	22,1	6,14	per room
Hotel low class	100	13,82	3,84	per room
Fitness centers	35	4,84	1,34	per user
Laundry	6	0,83	0,23	per kg
Restaurants	10	1,38	0,38	per meal
Bar	2	0,27	0,076	per consumption

It is not easy to design a solar plant. As distinct from determining the dimensions for some conventional plant taking into consideration the most difficult conditions, thus ensuring the safety of users, when determining the dimensions of a solar system, such criteria would be wrong; determining the size of a plant based on the most unfavorable conditions would lead to unimaginable over dimensioning, as the system would have to be designed for winter conditions as the most unfavorable ones, which would lead to an excess of unused energy in other months.

Figure 11.1 shows the variation of energy parameters related to the sanitary hot water production system.

Keeping in mind that in such systems the collector represents a basic cost, from the economic perspective decisions are different from the same decisions for a conventional plant.

In solar systems, the sun does not provide for complete required energy, but only a certain fraction, whereas the remaining energy is provided from some auxiliary conventional source.

Based on the above said, it is obvious that determining the dimensions of the solar plant is not only a technical issue, but at the same time technical and economic issue.

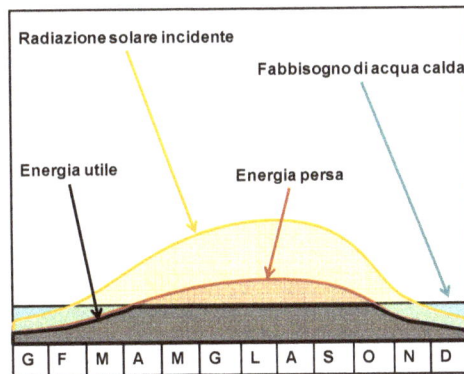

Figure 11.1 – Variation of energy parameters in a sanitary hot water production system

In technical terms, it is necessary to know the annual fraction of thermal load that can be met by a solar plant.

The fraction of useful collected energy depends on many parameters, primarily efficiency of solar collectors, which depends on the overall properties of the relevant collector, operating temperature, collector inclination and orientation, insulation, external temperature and wind velocity. Rational design of a solar plant requires a careful anticipation of the useful energy that could be provided by the plant, and such a calculation can be done using various methods and instruments.

There are various methods of calculation and the most widely spread is TRNSYS, or the simplest methods based on semi-empirical correlations, such as *f-Chart*, the description of which can be found in the References.

Although simplified, the *f-Chart* method provides sufficiently correct data, and it has been developed at the University of Wisconsin – Madison (USA) by the authors Klein, Beckmann and Duffie.

It has been obtained based on the results of numerous simulations of reference plants with the use of TRNSYS.

Using this method, it is possible to calculate the fraction f of the monthly thermal demand for hot water compensated by a solar plant.

The fraction f is defined as:

$$f = \frac{L-E}{L}$$

Where:

L = monthly thermal load;

E = auxiliary monthly energy.

If the monthly solar fraction f is known for all the months in the year, the annual fraction F is calculated in the following manner:

$$F = \frac{\sum f_i L_i}{\sum L_i}$$

Where:

f_i solar fraction of the month i;

L_i thermal demand for the month i.

In Figure 5.25, typical variation of annual solar fraction is shown depending on the collector area. For the given annual thermal demand L and for the given meteorological-climatic conditions, annual solar fraction increases logarithmically with the increase of the solar field dimensions.

With its increase, the price of auxiliary energy is reduced, but the price of the plant increases, which is usually proportional to the collector area.

In the same Figure, there is an analytical expression of the annual solar fraction depending on the overall solar field area.

The two coefficients c_1 and c_2 depend on thermal load, plant properties and the climatic profile of installation site.

Solar plants are very expensive investments due to the collector area, but the costs related to their functioning are low.

Larger areas are affected by higher solar radiation, but if a plant is dimensioned so as to cover the overall demand (f=1), as it has already been said, from the economic point of view it would not be cost-effective.

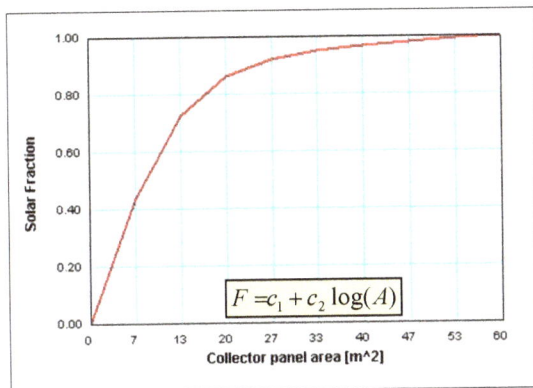

Figure 11.m – Variation of annual solar fraction in the function of the solar field area (typical graphical output of f-Chart software applied to the similar method).

Hence, it is necessary to design integrated solar plants where the fraction of the required thermal load is provided from the auxiliary source, usually powered by fuel oil or natural gas (methane), as already mentioned before.

Basic parameter in the design is the collector area A_c; with the increase of A_c, solar energy and energy saving are also increased since the energy from auxiliary source is reduced, but obviously the plant price also increases.

There is an optimal value of A_c which minimizes the total price; that value can be established with the method of *Updated Global Price (UGP)*.

In that method, various items occur, which affect the total price of the plant. These items are:

- the price of collector, water heater, hydraulic circuit, supports, pumps etc. I represents the initial investment, and a significant portion of this investment is proportional to the collector area;
- the price of electricity consumption for the operation of pumps;
- the price of auxiliary energy.

Collector area that minimizes the total price can be determined graphically by designating UGP in the function of the collector area.

Typical variation of the total annual price depending on the collector area is shown in Figure 5.26, where the presence of minimum dependence on the given solar field area can be observed.

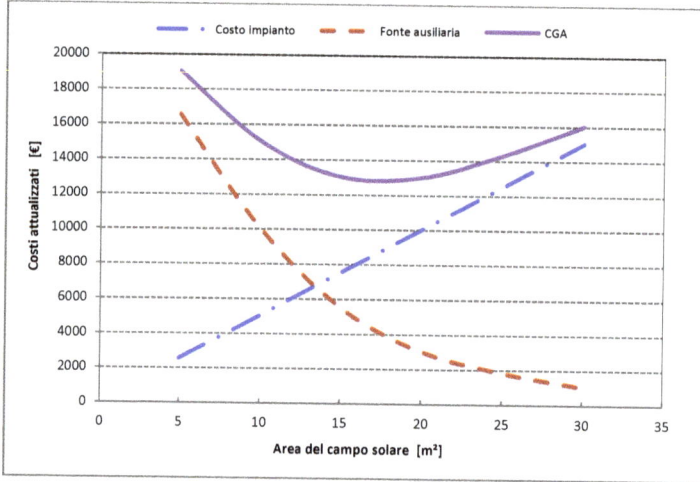

Figure 11.n – Variation of the total price of the plant depending on the collector area

11.6) Technical-economic data

Reduction of the price of thermal energy is a key for recognition of solar technologies, both in the low temperature applications for sanitary hot water production and space air-conditioning, and in the medium temperature applications for the production of heat for the processes in industrial use.

Investment and productivity of a solar plant are the basic factors determining the price of the produced heat energy.

For the purpose of reducing the kWh value of the heat energy produced by solar means, it is necessary to reduce the investment for plant itself, reduce the production price of the collector and its components, as well as to invest in the improvement of energy performances for the existing commercial systems.

Energy efficiency of a solar plant ranges between 600 and 800 kWh/m² depending on geographic location and the type of used collector.

In case of a plant for sanitary hot water production, the price varies depending on the desired amount of water, complexity of the plant installation, meteorological-climatic conditions at the installation site and type of integration (electrical, methane, fuel oil etc.).

These factors complicate technical-economic analysis, in any case for the system intended only for sanitary hot water production for small consumers, key indicative costs range between 800 and 1200 €/m².

These prices are reduced progressively with the increase of system size.

With these expenses and without incentive measures it is not easy to achieve economic sustainability.

In comparison with heat energy production with the use of gas boilers or electric water heaters, average annual production of about 700 kWh per m² of installed solar collector corresponds to the saving per square meter of about 60-70 €/annually for the unused consumption of gas from boilers and about almost double value (130-140 €/m² annually) in case of replacing electric water heaters.

Without incentive measures, investment return period is about 6-7 years in case of the replacement of electric water heaters and about 13-14 years in case of gas boilers. These periods almost have reduced by half owing to the current de-fiscalization of 55%.

11.7) Room air-conditioning

Solar systems have been developed and improved over time in respect of their quality. Rather high quality standard has been achieved and advantage with regard to the efficiency and overall operation of solar systems, starting from the panels for thermal storage to the integral parts.

National and international market offers a wide choice depending on application and the use of structure to which the solar system is applied.

The use of a renewable source such as solar energy can be interesting from economic point of view and from the perspective of technical feasibility if air-conditioning systems, both winter and summer type, are included in the series of works with an aim to realize buildings characterized by high energy efficiency.

Solar heating and cooling systems use solar energy for the purpose of heating and cooling in buildings and contribute to the consumption of energy obtained from fossil fuels i.e. reduce the emission of harmful gases and increase the percentage of utilization of renewable sources, as prescribed generally by the laws on the promotion of the use of energy from renewable sources.

This law prescribes that in new buildings, or in the buildings which are to be considerably renovated, heat energy production systems shall be designed and constructed so as to ensure covering 50% of anticipated consumption of sanitary hot water from renewable sources and the following percentages of overall consumption for sanitary hot water, heating and cooling:

- 20% in case the application is submitted from 31 May 2012 to 31 December 2013;
- 35% in case the application is submitted from 1 January 2014 to 31 December 2016;
- 50% in case the application is submitted after 1 January 2017.

Direct use of solar energy is even more cost-efficient when it is used for sanitary hot water production, heating and cooling of rooms. Solar heating and cooling systems use solar energy during the whole year thus reducing the system depreciation period.

Room heating by means of solar sources has recorded increase with the introduction of low temperature heating systems.

Low temperature heating systems – such as floor system or system installed in ceiling or walls and, partially, fan coil units - have a great advantage of operation at supply temperatures ranging from about 30 °C to 45-50 °C.

Energy crisis during the 1970s showed that solutions with lower energy consumption were required, so that the use of heating system terminals, which could heat the rooms by means of heat-conveying fluid at lower temperatures, was specifically taken into consideration.

Among these components, panels are distinguished, whose thermal inertia is better than in other types of terminals (constant and high), and which ensure higher thermal energy than energy generated in other systems.

Energy issues have led to the implementation of laws that also govern the insulation of buildings. It becomes possible to reduce their existing thermal power and their thermal level using lower temperatures.

Namely, the medium level of insulation for a building allows heating of apartments with the use of lower temperatures.

This progress encouraged the implementation of increasingly reliable systems characterized by high thermal efficiency and valid contribution to meeting the energy needs of a building.

The level of integration increases simultaneously with the technological progress of solar thermal systems.

There are two technologies in the market today that are in compliance with these applications:

- Flat solar collector;
- Vacuum tube solar collector.

In Figure 5.27, efficiency curves are shown for a flat-plate solar collector (continuous curve) and a vacuum tube solar collector (dot-dashed curve).

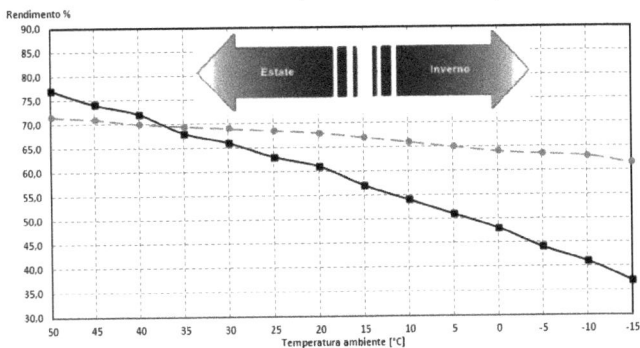

Figure 11.0 – Efficiency curve with radiation of 800 W/m² and average internal temperature of heat-conveying fluid of 50 °C

It is obvious that vacuum panels are steadier in maintaining the efficiency in case of fall of external temperature.

Room cooling by means of a solar source has become a valid alternative to conventional systems, particularly during the recent years, when electricity demand in the summer period has reached extremely high values due to excessive use of conventional air-conditioning devices, which in some cases has even led to power failure.

The use of solar energy for cooling has its advantages, which is shown by numerous pilot projects in other countries in Europe, primarily in Germany and Spain.

The use of solar energy for cooling of buildings is attractive since the highest demand for air-conditioning coincides with the months when solar radiation is the highest and when days are the longest.

Air-conditioning systems operated by solar energy have an undoubted advantage due to the use of non-harmful operating fluids, such as water and salt solution.

They respect the environment, meet the criteria of energy efficiency and can be used alone or integrated into conventional air-conditioning systems for the purpose of air quality improvement in any type of building.

Their basic aim is to use "zero emission" technologies in order to reduce energy consumption and CO_2 emission.

The general principle of these air-conditioning systems is production of cooling from heat sources.

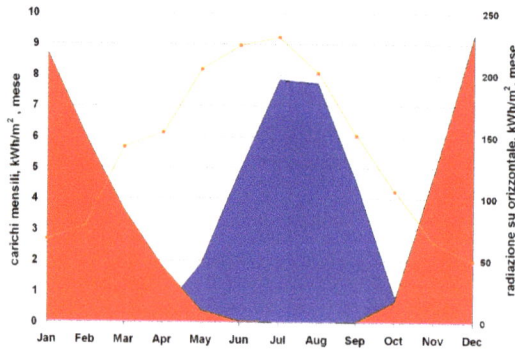

Figure 11.p – Variation of the demand for cooling energy

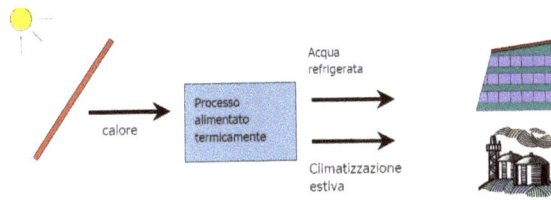

Figure 11.q – Operating principle of air-conditioning with the use of heat sources

For giving a brief view of the production system, one may underline:

- Heat source, the "sun," emits energy which is absorbed by solar collectors.
- Production of cooling is carried out with the use of refrigerating machines, which are supplied with hot water produced by solar collectors.
- Cold heat-conveying fluid, water or air, depending on the type of machine, is used for room air-conditioning.

11.8) Space solar-heating system

In the few recent years, space heating with solar systems has not been reserved only for the purpose of research, but it becomes pretty established reality.

This system ensures optimal operation, since the following conditions are adhered to:

- Low temperature heating system, with panels in floor/wall/ceiling or low/medium temperature system with the use of fan coil units;
- System of high efficient solar panels that can provide energy or a fraction of energy (along with integration) required for maintaining comfortable room temperature during harsh seasons;
- Storage of heat energy (which originates from a solar system, boiler, etc.);
- Interactive system, which ensures the production of heat in case there is no solar radiation.

In winter, with the use of solar-heating technology, solar field produce hot water at the place of traditional heat production system operated by fossil fuels; then the heat obtained in such a manner is stored in a storage tank from where it is sent to building, with a possibility to supply the circuit of sanitary hot water and heating circuit.

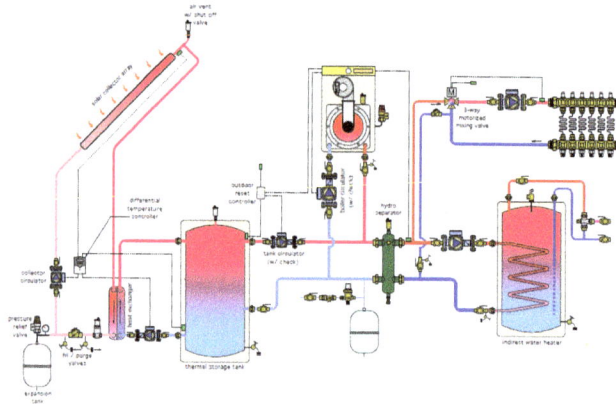

Figure 11.r – Operating diagram of solar heating

11.9) Low temperature heating systems

Solar source is suitable for connection to low temperature heating systems such as floor panels, since owing to their large area they require lower temperatures of heat-conveying fluid at the inlet.

Typically, these system terminals are operated by hot water, which is at the temperature of 40 °C at the inlet, thus enabling energy savings as compared to higher temperature systems.

The purpose of radiant systems is to distribute the energy for room heating and cooling.

They are primarily based on the exchange of heat for radiation through large exchange surfaces (hence the name "radiant systems").

For instance, radiators cannot be primary source of exchange, but a prerequisite, since high temperatures and areas are reduced at these terminals.

Radiant systems, however, can reckon on considerable areas, and do not require high temperatures at concentrated points which cause thermal inconveniencies.

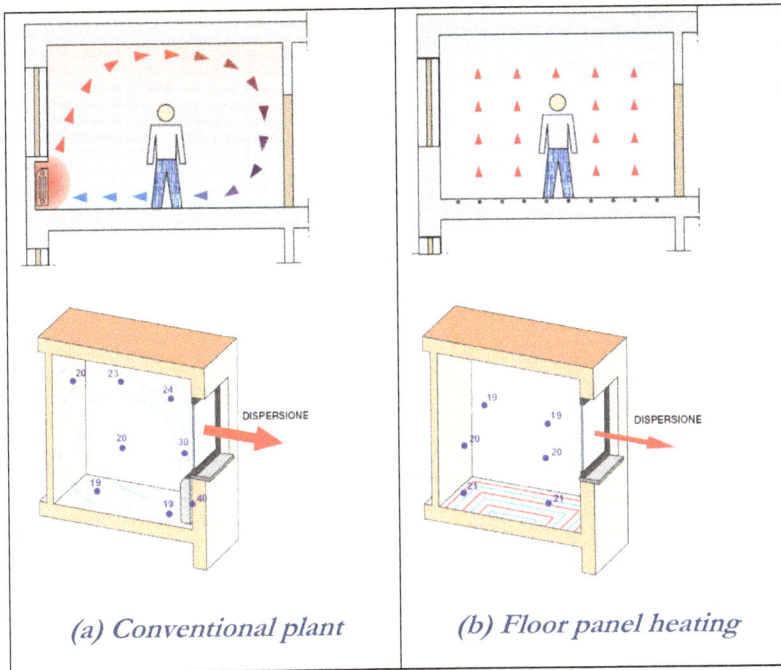

(a) Conventional plant *(b) Floor panel heating*

Figure 11.s – Comparison of conventional and floor system

A radiant system consists of pipes, usually made of plastic material, which cannot be seen, and are located within building systems (floor, wall, ceiling), and it is distributed so as to reach uniformly all the amount on the surface on which it is installed.

The heat is conveyed by pipes located in the floor that becomes a radiant area and uniformly distributed system.

This property of surface uniformity allows the use of low temperature fluid in the pipes, or more exactly, it has a reduced ΔT as compared to external temperature.

This thermal difference is reduced both in the winter and in the summer. In contrast to radiant systems, conventional radiators placed at some specific location in the room require temperature of 70 - 80 °C to be able to heat all parts of an apartment.

Namely, the heat in this case comes from small exchanging areas.

This property does not ensure ideal thermal comfort and causes radiant asymmetries that may result in inconveniences to people in the room.

Radiant systems have the following advantages:

- Energy saving due to lower temperatures (heat pumps and condensing boilers can be used).
- Lower operating costs due to lower energy consumption.
- Thermal comfort due to better temperature distribution in the room (higher temperatures in lower parts).
- Lower ΔT reduces dust fall in the rooms due to convective motion.
- Lower room temperature makes breathing easier for people in the room.
- Absence of convection phenomena due to heating and dust fall.
- Esthetic advantage since heating bodies or interior heating units are removed.
- Practical advantages since heating bodies or interior heating units, which occupy the space, are removed.
- Silence as water slowly flows through the pipes.
- Longer duration and less need for maintenance if installed with quality and in accordance with code of code of conduct.
- Supply to radiant panels with the use of solar thermal systems owing to the required low temperatures (about 40 °C).
- Pipes in heating networks need not to be installed on perimeter building walls so that the problem of water freezing is avoided when the system is not in function.

Disadvantages of a radiant system are the following:

- High thermal inertia (constant in time from 4 to 10 hours depending on the type) and less flexible and more complex control. This problem can be overcome by low screed or modern dry wall systems which include a concrete screed.
- Careful design.
- System with more complications in the existing buildings. Systems are simpler in newly erected buildings.
- If conventional systems are already present, e.g. radiators, heating plant has to provide water in the network with two different thermal levels, one lower for radiant systems and the other, higher, for conventional systems.
- Several centimeters greater thickness of the floor.
- It is necessary to control humidity in other systems (dehumidifiers, primary air system) in summer time.
- Higher initial costs.

Dati generali	
Temperatura di mandata	37°C
Temperatura di ritorno	32°C
Temperatura ambiente	20°C
Tubo	Mixal 16x2
Conducibilità del calcestruzzo	1.28 W/mK
Spessore del calcestruzzo	5 cm

Rivestimento ceramica 10 mm ($R_{\lambda,B}$ = 0,01 m²K/W)

Passo di posa [cm]	T	5	7,5	10	15	20	22,5	30	35
Potenza termica specifica [W/m²]	q	98	91	85	74	65	60	50	50
Temperatura superficiale [°C]	$\theta_{F,m}$	28,8	28,3	27,8	26,8	26,0	25,7	24,7	24,7

Rivestimento cotto 15 mm ($R_{\lambda,B}$ = 0,0167 m²K/W)

Passo di posa [cm]	T	5	7,5	10	15	20	22,5	30	35
Potenza termica specifica [W/m²]	q	94	87	81	71	62	58	48	42
Temperatura superficiale [°C]	$\theta_{F,m}$	28,5	27,9	27,4	26,6	25,8	25,5	24,6	24,1

Rivestimento parquet 12 mm ($R_{\lambda,B}$ = 0,06 m²K/W)

Passo di posa [cm]	T	5	7,5	10	15	20	22,5	30	35
Potenza termica specifica [W/m²]	q	72	68	64	57	51	48	40	36
Temperatura superficiale [°C]	$\theta_{F,m}$	26,6	26,3	26,0	25,4	24,8	24,6	23,9	23,6

Rivestimento moquette 10 mm ($R_{\lambda,B}$ = 0,11 m²K/W)

Passo di posa [cm]	T	5	7,5	10	15	20	22,5	30	35
Potenza termica specifica [W/m²]	q	57	54	51	46	42	40	35	32
Temperatura superficiale [°C]	$\theta_{F,m}$	25,4	25,1	24,9	24,5	24,1	23,9	23,4	23,2

Figure 11.t – Values of thermal efficiency calculated based on the values specified in the general data table

Regulations that govern the definitions, calculation of the system emission power, general criteria for determining the dimensions and installation, selection of the system components for floor heating are the regulations UNI EN 1264 consisting of four parts.

Maximum floor temperatures are:

- 29 °C in the living room;
- 33 °C in the bathroom and similar rooms;
- 35 °C in peripheral rooms.

With the use of maximum floor temperatures, values of thermal efficiency are shown under the specified conditions of general data.

System control logic in winter operation

These systems require well defined control logic in order to function with the highest efficiency.

During the day, in the presence of radiation, electronic central unit, with appropriate temperature probes and necessary automatic elements, has a function to automatically control the whole heating system.

The central control unit, based on the thermal probe signal on the solar field, activates circulation of heat-conveying fluid in the primary circuit and conveys energy to the thermal accumulator of combined type, particularly in low temperature heating systems, with an aim to ensure heating and sanitary hot water production.

Central unit, through various probes placed above the boiling vessel (accumulator) controls stratification and manages integration via traditional heat generation systems (e.g. boilers or pumps).

System control is basic, since if temperature is established which is insufficient for the consumer, it provides consent for start-up/shut-down of the solar system and thermal machine which is to bring the flywheel to the desired temperature.

Usually low temperature heating systems are to include a central storage part in the service of heating system, and high part for sanitary hot water production.

Therefore, there is tendency to provide a boiler or pump in two different methods and on two parts of the flywheel:

- in the central part with lower temperatures. This aspect is very important since excessively high temperatures may cause lack of differential between flywheel and solar field, thus limiting energy transfer;
- in the higher part with higher temperatures, which are adequate for sanitary hot water production.

Therefore, it can be stated that the flywheel and temperature control are fundamental aspects for proper operation of the solar system, particularly in winter months when the contribution of sun is limited and when it is important to correctly store the greatest possible amount of energy.

11.10) Space solar-cooling system

High increase of demand for air-conditioning in buildings, for which a constant increase is also anticipated in the following decades, has caused, as it has already been said, large electricity consumption in the summer period: until now, for summer cooling, air-conditioning devices with compression refrigeration machines have been primarily used, whose compressor is operated by an engine that absorbs electricity; this is a basic cause of the peak electric power required in the summer period, and which in numerous cases reaches the limits of the grid capacity.

Discharges of gases that cause greenhouse effect are increased with energy generation from fossil fuels and with the use of cooling fluids that change the climate and additionally aggravate the chain process which is the cause of all changes.

This situation is also confirmed by the expansion of air-conditioning devices in the market: for less than ten years, the number of products has increased more than five times.

In such a situation, new concepts of buildings become interesting, which on one hand tend to reduce cooling by implementing passive and innovative measures and, on the other hand, to use alternative solutions for covering the remaining demand for cooling.

Alternative air-conditioning devices are primarily the devices operated on solar energy, which ensure reduced electricity consumption, and which can use solar energy available right in the period of the greatest demand for air-conditioning.

The systems convert thermal contribution into the cooling output and can be classified in two types: systems with open cycle and systems with closed cycle.

Systems with open cycle (Figure 5.33) use cooling water for direct air treatment; hence, a distribution network for cooling is always required, based on the ventilation system.

Closed systems (Figure 5.34), however, consist of refrigerating machines, which are supplied through thermal conductors, hot water or steam, and which produce cool water; heat-conveying fluid can be used directly in treatment units in air-conditioning devices (cooling or dehumidification in the device) or can be distributed through a piping system to air-conditioning terminals decentralized in various locations that are to be air-conditioned.

They can be used with any technology for cold distribution (ventilation systems, fan coils, radiant areas).

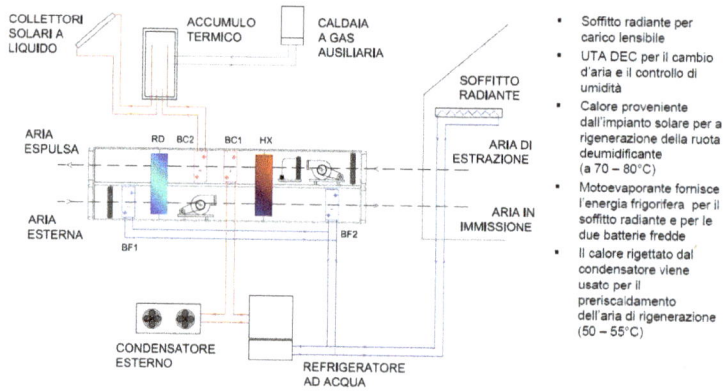

Figure 11.u – Open system combined with a radiant system supplied by a refrigeration compression group

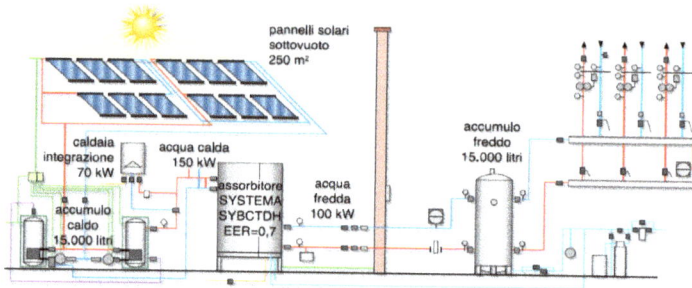

Figure 11.v - Closed system. Solar cooling system with an integrated boiler and storage system for hot and cold water. Indicative hydraulic diagram

In this segment, there are two types of refrigerating machines:

- Absorption machines, covering about 80% of the market.
- Adsorption machines, with several hundred applications around the world, but with an increasing interest for solar energy applications.

For many years (more precisely, since the end of 19th century) absorption coolers on hot water – steam – gas, have been used, but usually for high powers.

In the recent years, absorption refrigeration technology has advanced with the use of new materials and electronic components, and has increased the qualitative standard of machines with regard to efficiency and reliability.

There is a particularly strong development of low power absorption refrigerating machines that are supplied directly with hot water (up to 90 °C) or superheated (180 °C) which, as it can be easily assumed, opens an interesting perspective for the use in solar collectors.

11.11) Absorption cooling groups and solar-cooling

At the time that is increasingly directed towards compensation of raw materials and energy, absorption cooling groups are quickly spread as they can produce cooling with the use of thermal waste from industrial plants, unused heat in cogeneration systems or free heat (such as sun's heat), requiring no mechanical energy.

This cooling production system becomes an alternative to the compression refrigerating machines; in contrast to them, where the cooling effect is obtained from mechanical energy, absorption machines produce cooling with the use of thermal energy.

Operating principle of an absorption machine is most frequently based on the properties of the solution H_2O-LiBr, or on the relationship between cooling water and lithium bromide absorber, and on the fact that for the same temperature, the pressure of water vapor above the liquid is less than the pressure of vapor saturated with water.

In the most frequent version (simple absorption cycle with one stage) absorption machine is a three-temperature system operating at two pressures.

Three-temperature system operates at three temperatures: in conventional absorption machine whose cycle is described, it is about the temperature of produced cooling, temperature of heat transfer to the environment and higher temperatures of thermal input to the machine.

Simple but suggestive representation of the cycle can be obtained in the diagram temperature/pressure, where the temperature and pressure levels can be recognized for each of the four machine blocks (Figure 11.z).

The coolant evaporates and takes away the heat from the cold source at the temperature T_0, and at lower pressure p_0; cooling vapor is absorbed at almost the same pressure at medium temperature T_1 in the absorber of the solution absorbent-coolant. The solution is regenerated at the temperature and pressure p_1 higher than the generator making room for cooling vapor which condenses at medium temperature and pressure p_1 of the condenser, while weak solution (in the cooler) returns to the absorber at medium temperature and lower pressure.

Evaporator provides useful refrigerating effect, i.e. takes away the heat energy at the temperature T_0 from the cold source, while the heat energy at higher temperature of the cycle T_g is conveyed to generator.

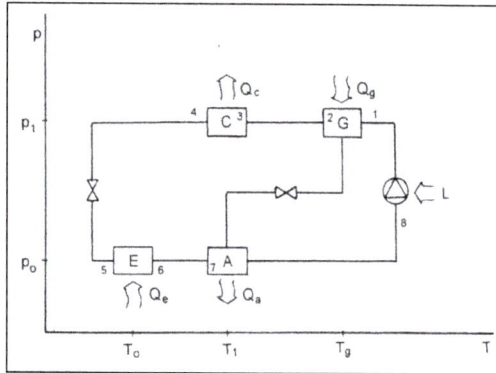

Figure 11.z – Block diagram of the temperature-pressure plan for the machine with elementary absorption

Absorber and condenser are to be adequately cooled at medium temperature. Transition of the fluid from higher to lower pressure (from the condenser to evaporator and from generator to absorber) is prevented with the use of lamination Transition of the fluid from lower to higher pressure (from the absorber to generator) obviously required energy: as the solution is liquid, a pump that does not consume a lot of energy will be adequate.

Hot water at the inlet conveys heat to the generator (about 90 °C), while the condenser and absorber convey heat to cooling water which comes out, heated at low temperature (about 35 °C), whereas the evaporator produces cooling water (temperature usually about 7 °C).

This cold fluid represents the output of the system.

Figure 5.36 shows a typical diagram of an absorption refrigerating machine.

It is obvious that due to the effect of the heat Q_E coming from outside, water in the evaporator evaporates and is absorbed by the recipient (absorber which contains solution where a thermal flow is generated towards the external Q_A. This solution, which is rich with water, is pumped in the generator where, with the use of Q_G, it is reduced and led to the initial concentration without water vapor, i.e. it is sent to the absorber.

Water steam extracted in generator is sent to the condenser where it is condensed and Q_C is given away. Observing the relationship between the generator and absorber, it can be noticed that there is a transition from hot concentrated solution (absorber) that is transferred to the absorber, and the transfer of diluted solution (coolant and i absorbent) in generator.

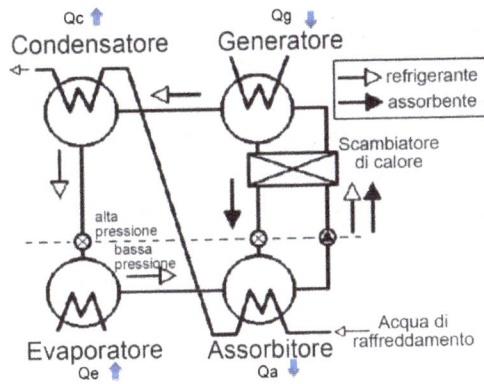

Figure 11.w – Diagram of an absorption refrigerating machine complex (with a heat exchanger between the fluids moving between absorber and generator)

By including a heat exchanger between the two flows, it is possible to reduce the temperature of concentrated solution entering the absorber, along with the increase of the absorption capacity and to increase the temperature of diluted solution entering the generator, decreasing the amount of heat that is to be provided to the system.

For conveying the heat in the condenser, cooling liquid and the heat generated in the absorber, it is necessary to have separate cooling circuits.

A larger portion of heat generated in the absorber is the condensation heat of the cooling fluid; approximately the same quantity has to be reduced from the condenser. Based on the pressure difference in these two components, it is concluded that the temperature in the condenser is higher than the temperature in the absorber.

There can also be only one cooling circuit with the inlet to the absorber and the outlet from the condenser.

The same as in each type of refrigerating machine, cooling liquid should have high evaporation heat and should be stable, so as not to be replaced after a certain time, and not to become corrosive, inflammable and toxic.

The pairs frequently used in absorption machines are water- lithium bromide (H_2O-LiBr) and ammonia-water (NH_3-H_2O).

In the pair H_2O-LiBr, the coolant is l'H_2O (water), while in the pair NH_3-H_2O the coolant is NH_3 (ammonia).

This substance is more sensitive than water since it is:

- unstable/quickly evaporates, so it is necessary to release water-ammonia mixture in the generator and divide it into two components;
- toxic and non-inflammable;
- requires high condensation pressure (>10 bar).

Therefore, the pair NH_3-H_2O is less spread. However, its advantage is the fact that it is able to cool at the temperatures lower than 0 °C and uses a cooling circuit operated by air.

In the pair H_2O-LiBr, cooling fluid has a series of advantages:

- it is stable;
- it is not toxic;
- it is easily available;
- it has high latent evaporation heat.

However, it cannot go under 0 °C. Absorption substance that acts as a solution is lithium bromide having the following properties:

- it is a crystallized salt;
- high affinity to water;
- high boiling point;
- it is not toxic.

Application of an absorption cooling group with low temperature hot water produced in solar plants has many advantages:

- Operating the machines with the use of heat energy from hot water results in drastic energy savings. The groups can be compared to actual energy recuperators and can use legal reliefs.
- Reduced electricity consumption: primary energy is thermal, while electricity is used only for the operation of auxiliary control devices, circulation of operating fluids, and taking away the heat. For example, absorber with cooling 70 kW requires only 0,6 kW electric.

- Outdoor system: the units are designed to be installed even in the aggressive environments (atmosphere), such as in ports or industrial zones.

  Technical rooms for placing the cooling central unit are no more required. In case of a covered system, due to reduced size, the machine requires minimum space.

- High reliability of the absorption groups due to limited number of mechanically movable parts. This results in a lower need for maintenance owing to smaller number of components, and thus their control and replacement.

- Silence and duration: the groups are perfectly sound proof, very silent and vibration free.

  They are suitable for the installation in city centers in solar pavements, terraces, attics and construction yards intended for commercial, industrial and service activities. Since there is no friction and mechanical abrasion, machine life cycle is longer.

- Limited harmful effect on the environment: the groups do not use harmful fluids. Supply energy limits harmful emissions.

- Modular partitioning and control: in order to achieve higher efficiency of the absorber, it is possible to apply stage control on modular groups (more than one unit), partitioning their operation depending on the existing thermal load.

Despite the undoubted possibilities of the absorber, economic aspects, which aggravate their application, should not be ignored.

In the first place, high price of the absorber: this aspect is related to the reduced economy of scales as the market is still underdeveloped.

The increase in their application could result in price reduction.

11.12) Storage of thermal and cooling energy

In a solar cooling plant, thermal accumulator has a fundamental role in the plant efficiency, as well as its type and size in relation to the total collector surface.

For the continuous plant operation for the purpose of solar cooling, it is necessary to properly determine the size of the accumulator volume.

Heat storage is very important for separation of thermal energy from its use in a device for solar cooling. In a solar refrigeration system, it is possible to find a hot and cold side.

The purpose of the storage (accumulating) at the hot side is to store the heat produced in solar collectors and to transfer that heat to refrigerating machine generator when required.

By storage in cold water, cold produced in refrigerating machine is accumulated when solar energy is available, and subsequently transferred to the cold diffusion system, upon request. In Europe, heat storage at the hot side is more widely spread then cold storage, it is enough to think of the fact that in 33 plants analyzed by IEA, only in 19 heat storages at the cold side is applied, while everyone use storage at the hot side.

Just as in case of solar collectors and heat storage, for the same cooling power it varies a lot from plant to and from one country to another.

11.13) Heat storage

The same as cold side storage, the hot side storage is necessary for facing the imbalance between solar input and the needs for heating and cooling, so that solar energy is stored and can meet the demands for the production of cold (and heat if a solar cooling device is also used during the winter period for heating) despite the variability of needs.

To that effect, the higher spread of hot side storage is emphasized in solar cooling systems as compared to cold side storage, with regard to the fact that while these latter plants are used only in the summer period (or when there is a need for cooling), hot side storage is used both in the winter and summer.

An interesting application of hot storage is the application with a double storage tank, control valves in order to face the load differences during the same day and enable high capacity storage in the periods of maximum demand, as well as higher efficiency in the periods of lower demand (while avoiding to oversize the system).

In cases when it is assumed that accumulated thermal energy is not sufficient to meet the needs during the periods of the highest consumption an auxiliary source can be installed.

If the auxiliary source is placed in series with the storage, it allows the increase of water temperature at the storage outlet; in that case, it is said that it acts as a booster).

If the auxiliary source is placed in parallel, then it can fully cover thermal demand in case that the storage is not enough.

The first solution appears to be a reference solution considering that the purpose of solar cooling is reduction of the use of conventional energy sources.

However, this is not always the case since sometimes it is difficult to obtain a limited increase of water temperature at the storage outlet, unless the auxiliary source has a separate control system.

11.14) Cold storage

The purpose of cold storage is to shorten the cold production time by means of refrigerating machine and its use in air-conditioning.

This time difference can range from several hours to several weeks, and even months, depending on if it is a daily or seasonal storage.

The use of cold storage for solar cooling devices is conditioned by the need to solve the problems of non-continuous operation of the absorption machine.

Besides its positive effect on the machine performance, cold storage allows the implementation of devices with lower installed cooling power.

It is interesting to point out that inefficiency in daily storage actually does not necessarily imply waste of energy since, for example, higher operating costs can be obtained due to more regular operation of refrigerating machines with lower need for modulation.

A study of the authors Bo He and F. Setterwall is especially interesting, where a conclusion is made about the basic advantages of cold storage:

- Reduction of the operating costs of the plant.
- Fewer interruptions in the functioning of the absorption machine.
- Reduction of cooling equipment.
- Increase of operating flexibility.

11.15) Short-term heat storage

In short–term storage there are (Fig. 11k):

- Sensitive storage;
- Storage and phase change, in literature known as latent storage or PCM (*Phase Change Materials*).

In sensitive storage, thermal energy is accumulated raising the material temperature; therefore, the amount of stored heat is calculated in the following way:

$$Q = \int_{T_i}^{T_f} mc_p \Delta T$$

Where:

- ✓ T_i = initial temperature of the substance;
- ✓ T_f = final temperature of the substance;
- ✓ m = mass of the substance;
- ✓ c_p = specific heat of the substance.

In the majority of cases, (in almost all cases of solar cooling) water is accumulated, both at the hot and at the cold side, owing to its cost-efficiency and thermal capacity.

In the phase change accumulators, the amount of heat stored in the accumulator is calculated in the following way:

$$Q = \int_{T_i}^{T_m} mc_p \, dT + ma_m \Delta h_m + \int_{T_m}^{T_f} mc_p \, dT$$

Where:

T_m = fusion temperature of the material;

a_m = fraction of fused material;

Δh_m = latent fusion heat of the material per mass unit.

In contrast to sensitive accumulators, PCM accumulators during the phase change absorb and release the heat with almost constant temperature, storing at the same temperature up to 14 times more heat than sensitive accumulators (apparently, the higher the latent heat, the more heat will be stored at the constant temperature).

As it can be seen in Figure 5.37, material phase change can be carried out with the use of the following change of condition: solid-solid, liquid-gaseous and solid-liquid.

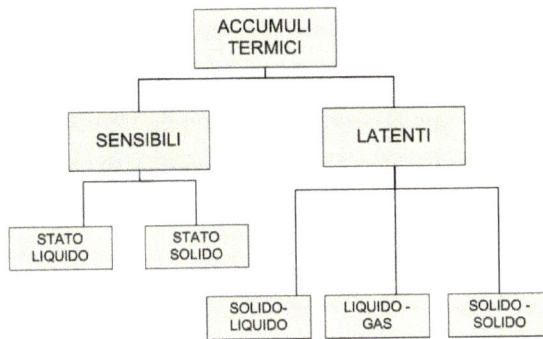

Figure 11.k – Classification of short-term thermal storage

In the transformations solid-solid, heat is accumulated while the substance transformation is carried out from one type of crystallization to another.

These transformations are usually characterized by lower latent heat as compared to the transformation solid-liquid.

Advantage in the use of this type of storage is greater ease of design and construction of the accumulators themselves.

The best materials are organic solutions of pentaerythritol, Li_2SO_4 and KHF_2.

The phase changes liquid-gas, gas-liquid allow high exchange of latent heat, but also large variations in quantity occurring during the phase transformation result in big problems in the control of accumulators.

These increased variations in quantity require the use of complex storage systems, and in accordance with that, accumulators with the phase change solid-liquid are usually chosen.

The phase changes solid-liquid, liquid-solid lead to the exchange of latent heat lower than in the transformation liquid-gas and still enable much lower variation in quantity (order of magnitude of 10% or less).

From the literature it can be concluded that this method of storage is the most cost-effective and the most frequent.

11.15) Coexistence of radiant panels and fan-coils

The spread of radiant systems has an increasing importance in air-conditioning. The summer use of these systems points to the criteria of temperature control of cooling water.

Temperature of a radiant system can be lowered below the temperature of dew in the air since the condensation of humid air component would damage the structures and materials.

Hence, for the appropriate control of humidity, dehumidification machines will be used.

Cooling by means of panels has two exact limitations: low efficiency of cooling and inability to dehumidify the air in the rooms (as distinct from fan coil units and split systems).

Cooling of air without dehumidification may result in excessive relative humidity (U.R.). Let us take the example of a room with air a: t = 32 °C; U.R. = 60%.

If we cool that air without dehumidification up to a: t = 26 °C its new relative humidity (which can be determined with the use of psychrometric chart) is:

➢ U.R. = 90%, which is totally unacceptable value, as U.R. may not exceed 65-70% in order to obtain valid thermal conditions.

For the control of humidity in schools, museums or other buildings having large area, conventional machines can be used for air treatment, with batteries for cooling and post-heating.

For houses and apartments, fan coil units can be used as humidity controllers (dehumidifiers).

Fan coil units (in order to be able to control air humidity) must use water at very low temperature (e.g. supply/return 7/12 °C).

In addition, their advantage is the ability to integrate cooling performances of the panels.

Fan coil units (or fan-coils) are the terminals that add heat to (or take it away from) the environment mostly through forced convection.

They consist of one or two finned batteries with an exchanger, one or two centrifugal fans (capable of speed selection), grate for air intake (on the bottom or front side) with a suitable filter, vessels for collecting the condensate (which is generated due to the condition: $T_{supbatt} < T_{rosa}$) and housing (made of varnished sheet metal or plastic material).

This type of unit, owing to the forced thermal exchange, allows obtaining of summer cooling by circulation of cooling air in the finned battery.

In this situation, humid air comes into contact with the battery fins and frequently has the temperature lower than the dew temperature, so that water vapor is condensed and has to be collected and removed.

Fan coil units are the terminals of the devices that have a modest thermal inertia: this means that in the rooms where they are located, desired temperature cannot be quickly achieved, but they quickly return to the initial temperature.

Hence, they are used in buildings where they are needed for a certain time during the day (office buildings).

When determining the dimensions of these bodies, after it is confirmed that the power in the conditions of functioning is adequate for thermal load, it is necessary to control the supply of the delivered air, as well.

This amount, measured in a number of re-circulations of the room air, represents an important design parameter because it primarily depends on the temperature uniformity.

Minimum recommended value is 5 re-circulations per hour: lower values will not ensure uniform temperature distribution, while excessive values can result in excessive air speed with a negative effect on the health.

11.17) Plant control logic during summer operation

In a solar thermal system for room cooling, besides the above described solar systems, there must be a control logic that controls all the devices in the field (electric pumps, electric valves, absorption cooling group etc.) depending on the value of temperature achieved with a controller.

Production and storage of hot water generated in a solar field will be independent of the demand for cooling energy which is required for the air-conditioning of the building: the aim is to maximize the production of heat energy obtained by means of solar field, making it available for supplying the generator of absorption cooling group. This latter element will carry the cooled water storage tank (Fig. 11.x).

Figure 11.x – Operational logic of the solar cooling plant

An air-conditioned building, depending on temperature in various environments, is served by cold accumulators previously loaded by a cooling group.

In case the building does not require refrigerated water, but a hot accumulator is still at the optimal temperature for supplying the cooling group, it will function and produce refrigerated water that will be stored in the cold accumulator, ready to be used when required for the building.

12. CSP: concentrating solar power

The so-called CSP technologies (Concentrating Solar Power Technologies) primarily include concentrated solar thermal power plants (also called thermoelectric systems) and chemical solar power plants: the first one is already used for electrical power generation, while the second one, still under development, will be used for producing hydrogen from water through thermo-chemical cycles, in order to generate energy.

Thermodynamic solar technology uses high temperature heat, produced by concentrating solar energy via optical systems (parabolic mirrors, Fig. 12.a), to generate electrical power using thermodynamic cycles (Rankine cycle, Stirling cycle etc.) in the same manner, as in conventional thermal power plants.

Solar thermodynamic plants are similar to thermoelectric power plants, whereby the fuel tanks are replaced by a solar field (a set of solar collectors), which means that the fossil fuel is replaced by a solar source, which is free and unlimited.

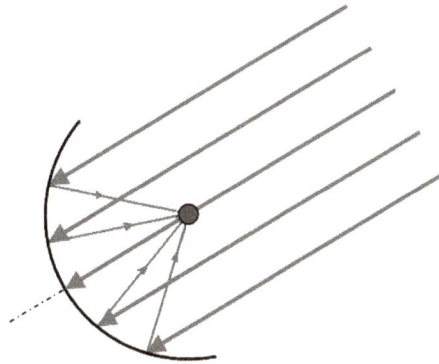

Figure 12.a – Concentrating solar irradiance using parabolic mirrors in a thermodynamic solar power plant

Concentrated solar power (also called concentrating solar power, concentrated solar thermal and CSP) systems use mirrors or lenses to concentrate a large area of sunlight, or solar thermal energy, onto a small area.

Electrical power is produced when the concentrated light is converted to heat, which drives a heat engine (usually a steam turbine) connected to an electrical power generator or powers an, experimental as of 2013, thermo-chemical reaction.

CSP is being widely commercialized and the CSP market has seen about 740 MW of generating capacity added between 2007 and the end of 2010.

More than half of this (about 478 MW) was installed during 2010, bringing the global total to 1095 MW.

Spain added 400 MW in 2010, taking the global lead with a total of 632 MW, while the US ended the year with 509 MW after adding 78 MW, including two fossil–CSP hybrid plants.

The Middle East is also ramping up their plans to install CSP based projects and as a part of that Plan, Shams-I the largest CSP Project in the world has been installed in Abu Dhabi, by MASDAR.

CSP growth is expected to continue at a fast pace. As of April 2011, another 946 MW of capacity was under construction in Spain with total new capacity of 1,789 MW expected to be in operation by the end of 2013.

A further 1.5 GW of parabolic-trough and power-tower plants were under construction in the US, and contracts signed for at least another 6.2 GW.

Interest is also notable in North Africa and the Middle East, as well as India and China.

The global market has been dominated by parabolic-trough plants, which account for 90% of CSP plants.

CSP plants can be broken down into two groups, based on whether the solar collectors concentrate the sun rays along a focal line or on a single focal point (with much higher concentration factors).

- *Line-focusing systems* include parabolic trough and linear Fresnel plants and have single-axis tracking systems.
- *Point-focusing systems* include solar dish systems and solar tower plants and include two-axis tracking systems to concentrate the power of the sun.

CSP is not to be confused with concentrated photovoltaics (CPV) we already saw.

In fact, in CPV technological process, the concentrated sunlight is converted directly to electricity via the photovoltaic effect of course.

Physical phenomena related to concentrating solar irradiance require that the radiation itself is characterized by a precise direction (direct radiation); thus, the CSP systems are not able to utilize the diffuse solar radiation and do not operate in conditions of overcast sky or thick fog.

In addition, a mechanical system is required, to continuously direct the mirrors towards the Sun, taking into account daily and seasonal variations.

The same system also has to ensure that the collectors may be rapidly diverted from the focus, in case the heat collector circuits are out of order, and to place them in a safe position when not in function due to a storm or strong wind.

Apart from generating electrical power through thermodynamic cycles, concentrated solar power plants are used for 'thermal' purposes, such as generating the heat for industrial processing or air-conditioning, but compared to a solar thermal power plant, it reaches higher temperatures, typically over 150 °C.

Compared to photovoltaic plants, concentrated solar thermal power plants have higher complexity, making them less appropriate for widespread use, but enabling generation of higher quantities of electric energy for the same installed power. Furthermore, they have similar operating principles as conventional plants using fossil fuels: due to their ability to accumulate thermal energy and to integrate easily with plants on fossil fuels, they may ensure a more stable and programmable service, allowing that, with certain restrictions, the power generation period be shifted with respect to the time when the solar irradiance is at its disposal, and to concentrate it in the periods of high demand, so that the energy achieves better prices.

From the viewpoint of the collector geometry, it is possible to identify, basically, three different technologies of thermodynamic plants:

- ✓ parabolic dishes,
- ✓ central towers,
- ✓ linear parabolic collectors (or parabolic trough collectors), plus the fourth technology, essentially a variant of the third:
- ✓ linear Fresnel collectors.

Solar plants with parabolic dishes use reflective panels (typically rotating parabola), tracking the path of the Sun through the mechanism moving along two axes and concentrating the solar irradiance onto a receiver fitted in the focus.

The receiver may consist of a Stirling engine with an electric generator.

The parabolic shape in revolution can be approximated as a set of spherical mirrors mounted on the supporting structure.

Central tower plants use an array of flat reflective panels (heliostats) rotating individually along two axes to track the Sun and concentrating the solar irradiance onto a single receiver mounted on the top of the tower.

Within the receiver there is a circulating fluid that transfers the heat.

The set of heliostats constitutes a huge parabolic surface and enables achieving enormously high levels of concentration factor.

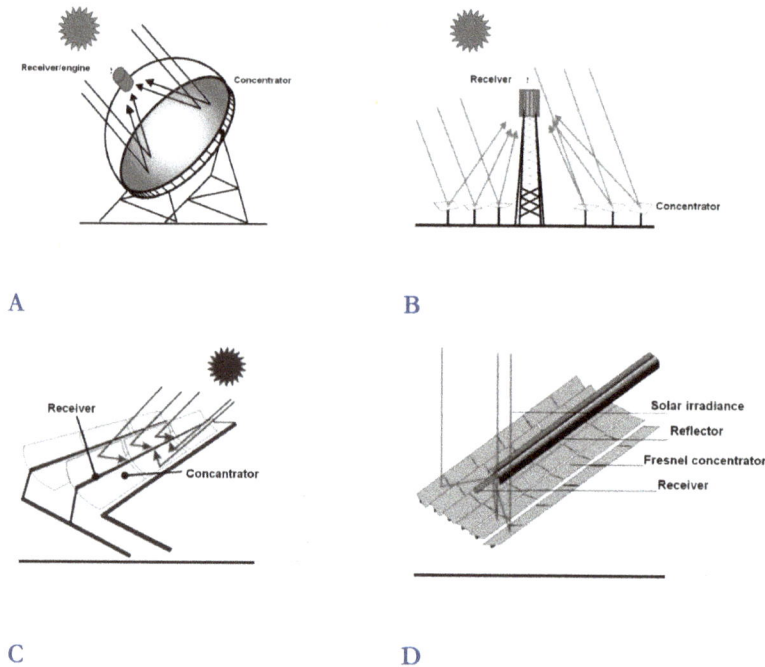

A   B

C   D

Figure 12.b – Technologies of solar thermodinamic plants: A) Parabolic dish. B) Central tower. C) Linear parabolic trough collector. D) Fresnel lens linear collectors.

The concentrator in the parabolic trough collectors has a surface in the shape of a translated parabola, or it may be described as an open cylinder with a parabolic base.

The collectors track the apparent path of the Sun using a rotating mechanism with only one axis.

Solar irradiance is focused in one axis (the focal point of each cross-section), where the receiving tube is mounted with the heat carrying fluid collecting thermal energy and transferring it to the electric power generation system or the accumulation system. Linear Fresnel collectors are a type of parabolic trough collectors: parabolic concentrator consists of flat mirror segments distributed by the Fresnel lens principle, whereby the receiving tube is placed into the focal point.

In this case, rotation required to track the Sun is related only to concentrator segments, while the receiving tube and the supporting structure remain stationary. Compared to parabolic trough collectors, this system has inferior performances (total annual output amounts around 60% of the output generated by parabolic trough collector), but the plant costs are lower since the receiving tube, not rotating together with the mirrors, does not require mobile joints (particularly expensive) and can easily use high pressure fluids, such as steam.

Additionally, mirror rotation system is simpler and the supporting structure requires less material.

The latest line of development relates to the so-called non-imaging systems, currently in the research phase, aiming to reach the theoretical concentration ratio 2-4 times higher than the one achieved in collectors based on conventional optics, with extremely wide receiving angles, in order to reduce and eventually eliminate the need for the continuous tracking the Sun.

One of the solutions under research is based on a cavity with reflective internal surfaces placed in such a manner that, through a series of reflections, solar irradiance entering the system is transferred to a small receiver placed at the bottom of the collector.

Various systems described above are in various stages of development: parabolic trough collectors have reached commercial maturity (even though there is still a lot of room for improvements), followed by central towers in a demonstrative phase, with a promising outlook for a full affirmation and a dish system, as well as the linear Fresnel collectors, at an experimental stage, offering interesting practical perspectives.

Non-imaging systems are still at the level of laboratory research.

Table 5.5 provides a summary of main characteristics of various systems.

Performances, in the sense of concentration factor, are higher in systems with two axis direction (central towers and parabolic dishes), with possibilities to reach higher temperatures and potentially higher efficiency in electric power generation.

CSP plants are typically industrial installations, similar to thermoelectric power plants with a power of an order of magnitude of at least several tens of MW, preferably several hundred MW, if one aims for better economies of scale, i.e. reducing the cost per unit installed power with the increase of the output power, primarily the conventional component of the plant (sector for power, control, etc...).

Table 12.c – Comparing the main solar technologies from the viewpoint of concentration

Technology	Direction	Concentration	Stage	Performance	Approx. price
Parabolic dishes	2 axes	up to 1000	Commercial	High	High
Central tower	2 axes	up to 1000	Demonstrative	High	Medium
Parabolic trough	1 axis	up to 100	Commercial	Medium	Medium
Linear Fresnel	1 axis	70 – 80	Pre-demonstrative	Low	Low
Non-imaging	1 or 2 axes	4000 (*)	Research	Very high	(*)

(*) Data assumed or unknown for the preliminary level of the technology

12.1) Parabolic trough collector technology

The parabolic trough collectors (PTC) consist of solar collectors (mirrors), heat receivers and support structures.

Parabolic trough collectors are currently the most widespread concentrated solar thermal power plants, accounting for over 90% installed power.

Regarding this type of plant, the typical configuration, shown in a diagram in Figure 512.d, includes the following parts:

a) Solar energy collection (solar field).

b) Steam production.

c) Thermodynamic converting of the heat energy to electrical and optionally.

d) Section for heat accumulation and integration (not shown in the figure).

Heat accumulation and integration allow providing a continuity of the service, compensating for the unpredictability of the primary source.

Solar field includes solar collectors that concentrate and reflect solar irradiance, receiving tubes that convert solar irradiance into thermal energy of high temperature and primary circuit for thermal energy collection and transport.

The solar field is the core part of the plant corresponding to the furnace of the thermoelectric power plant. In the solar field, collectors are aligned in parallel lines, each line consisting of several serially connected collectors.

Solar field has a modular structure: adding a series of collectors (modules) increases the collected thermal energy, and subsequently the power of the plant.

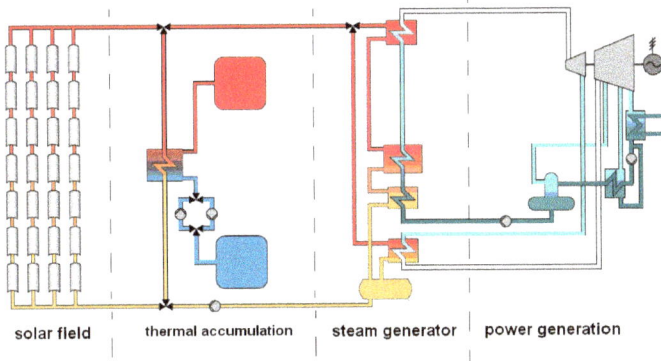

solar field   thermal accumulation   steam generator   power generation

Figure 12.e – Typical configuration of the concentrated solar thermal power plant with parabolic trough collectors

The collector arrangement in the solar field primarily depends on the site dynamics. Conventional arrangements consist of collectors oriented along the North-South or East-West direction, as a function of the site latitude and the planned manner of plant operation.

The tube is filled by the heat transfer fluid with a goal to collect and transfer the thermal energy; conventional solution is based on a primary circuit with thermal, mineral and synthetic oil, able to reach temperatures up to 400°C.

More innovative solutions use mixtures of molten salts that can reach higher temperatures (Table 12.f).

For producing steam, a heat exchanger ("solar" steam generator) is used, in which the thermal energy is transferred from the heat fluid to the pressurized water.

The steam generator may have a sector for overheating, producing the steam with thermodynamic properties adequate for sending to thermal power turbines (100 bar, 500 °C).

This solution requires the heat transfer fluid to reach temperatures higher than 500°C and cannot be used with thermal oil, which has the operational limit of less than 400°C.

There are two alternative possibilities, both posing a heavier load: using an auxiliary appliance for overheating, powered by fuel, e.g. methane, or using special turbines designed for various operating conditions.

The section serving for thermoelectric conversion uses steam to generate electric power, through water-steam thermal cycle, analogue to the cycle in conventional thermal power plants (Rankine cycle), including steam turbine connected to the power generator, condenser and the pre-heating systems for the supplied water.

This section includes auxiliary systems for condenser cooling, make-up water treatment, electrical panels, compressed air etc.

Table 12.f – Heat transfer fluids for concentrated solar thermal plants

Fluid	Temperature (°C) minimal	maximal	Critical aspect
Synthetic oil	13	395	Inflammable
Mineral oil	-10	300	Inflammable
Water/steam	0	>500	High pressure
Silicon oil	-40	400	Expensive
Nitrate salts	220	500	High solidifying temperature
Ion liquids	-75	416	Very expensive
Air	-183	>500	Low energy density

The heat accumulation sector is one of the most interesting characteristics of this type facility because it allows increasing the plant's functionality, i.e. number of operating hours per annum, with respect to the number of hours with solar irradiance at disposal, whereby electric power generation is more regular and easier to manage bearing in mind the variability of the solar source.

The most widespread solution is based on the system with two tanks and molten salt mixture, showed in a diagram in Figure 12.g.

At the accumulation stage, heat transfer fluid transfers part of the energy collected from the solar field to the salt mixture using the heat exchanger, whereby the heat is transferred from the 'cold' to the 'hot' tank.

At the stage of generation without salt, molten salts mixture returns the thermal energy to the heat transfer fluid over the heat exchanger, moving from the 'hot' to the 'cold' tank.

Heat accumulation capacity depends on the amount of salts molten in the tanks, i.e. of their size, and on over-dimensioning of the solar field with respect to the thermal power required by the power plant when operating under full load; when designing, an economic optimization is undertaken between plant cost increase and added value of production. Normally, an accumulation capacity corresponding to the period of 6-10 hours operation under full load is considered optimal.

As an alternative, or an addition to thermal accumulation, it is possible to integrate a solar plant with other energy sources.

This solution is interesting if biomass or other renewable sources are used as an integrative source.

Solar plant with parabolic trough collectors is focused on overcoming the restrictions imposed by utilization of heating oil for transferring the heat: maximal operating temperature is less than 400 °C and there are safety risks and environmental risks, since it is very inflammable and pollutes the environment.

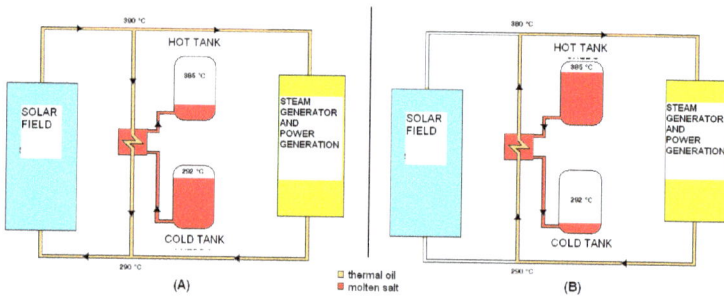

Figure 12.g – Heat accumulation operating principles: (A) generation and accumulation; (B) generation without the Sun

The new plant type uses molten salt mixtures, used not only for thermal accumulation, but also as a heat transfer fluid.

There are also systems utilizing a double fluid: thermal oil + molten salts (a recent example is Andasol, Spain, in operation since 2008), while 'Archimede' in Sicily is the only industrial-scale plant on the basis of molten salts, completed in 2010 (see Box 'Archimede Plant').

A very interesting characteristic of the concentrated solar power plants is that they allow integration into a conventional thermoelectric power plant, whose electric power generator they actually use (hybrid plants), sharing the personnel and all installations within the facility.

This opportunity is used in case of the Archimede plant.

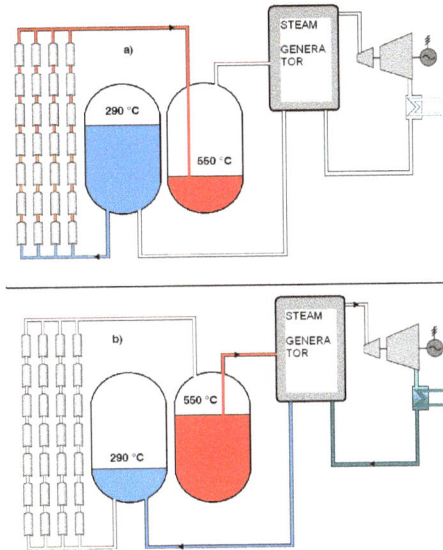

Figure 12.h – Diagram with the single fluid, molten salts: a) loading, b) discharging.

The Archimede Power Plant is the first industrial demonstration of the solar thermodynamic technology with parabolic trough collectors using molten salts.

Its nominal electric power is 5 MW and it is integrated in a thermoelectric power plant ENEL with a combined cycle, called Priolo Gargallo, in Sicily, with which it shares steam turbines and the thermal cycle, as well as the control room and the accompanying installations in the facility.

The project was initiated by the cooperation between ENEL, the big power italian Corporate and ENEA *(National Agency for New Technologies, Energy and Sustainable Economic Development)*, passing various phases of development.

In 2008, it entered the building phase, was finished in June 2010 and the power plant officially commenced operation on 14 July 2010.

It is managed by ENEL that takes care directly about operations and publishes the results obtained.

The heat collected in the Archimede's solar field is used to produce steam for turbines of the thermal power plant and allows generating additional electrical power from the solar source up to 9.2 million kWh per annum, which meets the household needs of 5000 persons, which further saves around 3000 toe annually and reducing the carbon dioxide emission of 5,500 tons a year.

The plant has an accumulation sector which consists of two tanks with different temperatures (550 and 290 °C) with over 1,500 tons of molten salts, equivalent to heat accumulation capacity of 6.5 hours.

Apart from the limited contribution to electric power generation, which is not negligible, the aim of the Archimede plant is primarily to demonstrate the industrial viability of thermodynamic solar technology of an innovative type, developed by ENEA, especially:

- to check the full functionality of the system over time;
- to check the reliability of various components;
- to optimize the operational procedures;
- to optimize the non-routine situations;
- to acquire the data for modeling physical processes involved;
- to collect experiences for further applications of this concept;
- to support developing component variants.

Operative experience of the Archimede plant is fundamentally important to develop applications on a larger scale and in more challenging environmental conditions, such as deserts and countries without adequate supporting technology.

While expecting further development, a plant sector can be devoted to experimental research of possible changes derived from the result obtained.

Based on the experience in designing and building Archimede plant, at the beginning of 2012, ENEL started a new project for building another solar plant in Sicily, of 30 MW; the initiative is conducted with a partial support of the EU and involves also ENEA and other European partners.

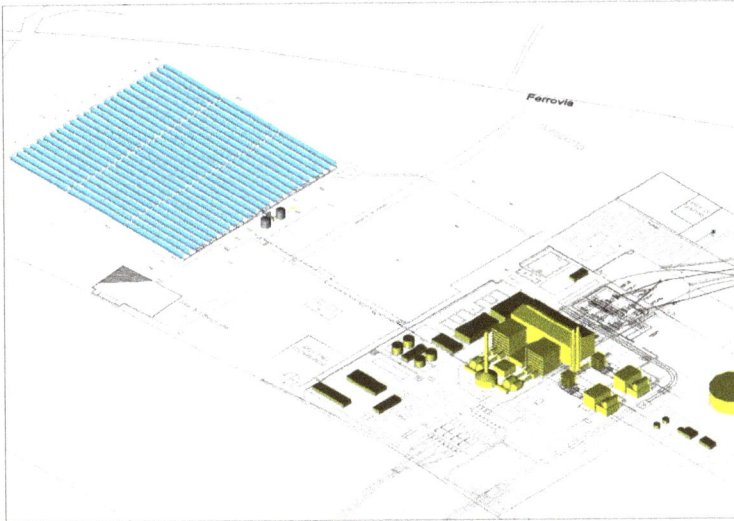

Figure 12.i - Solar power plant Archimede - graphical processing

The main data related to the plant are provided in the following table 12.l.

Table 12.1 - Design data of the solar plant Archimede

Site	Priolo Gargallo (Sicily)	
Area occupied	8	Hectares
Projected solar irradiance (DNI)	1936	kWh/m²/year
Expected annual generation	9.2	GWh/year
Technology	ENEA	
Design	ENEL-ENEA	
Ownership	ENEL	
Management	ENEL	
Beginning of the works	21 July 2008	
Start of the operations	14 July 2010	
Solar field area	3.860	m²
Number of collectors	54	
Number of loops	9	
Collectors per loop	6	
Collector total area	590	m²
Collector length	100	m
Number of modules per collector	8	
Collector manufacturer	COMES [4]	
Reflective panel manufacturer	Ronda Reflex	
Number of collector tubes	1,296	
Collector tubes manufacturer	Archimede Solar Energy [4]	
Heat transfer fluid	Molten salts mixture [1]	
Solar field input temperature	290	°C
Solar field output temperature	550	°C
Gross electrical power	5.0	MW
Net electrical power	4.72	MW
Turbine manufacturer	Tosi	
Power cycle pressure	93.83	Bar
Cooling	Seawater	
Turbine efficiency [2]	39.3	%
Gross annual efficiency [3]	15.6	%
Heat accumulation type	sensitive to heat with 2tanks	
Accumulation capacity	80 [5]	MWh term.
Accumulation medium	Molted salts mixture [1]	
Tank content	1.580	t
Tank height	6.5	m
Tank diameter	13.5	m

[1] 60% sodium nitrate, 40% potassium nitrate
[2] Full load
[3] Collected solar energy /Generated electric energy
[4] Under the ENEA license
[5] 6.5 hours under full load

Cogenerative applications also seem interesting, as well as heat generation for technological uses, which may enhance the field of application of these technologies to a medium-small volume, including localizations in areas which are not particularly convenient regarding insolation.

A promising solution for the near future is direct generation of steam in solar receivers, without intermediate fluids, in a manner that reduces the number of plant components, enhancing efficiency.

This option is under research in CIEMAT (Center for Energy Research of Spain) and DLR (Aero-spatial Center in Germany), conducting experimental tests on the Solar Platform in Almeria, Spain.

Another solution at the experimental stage is using a gas (e.g. CO_2) as a heat transfer fluid in solar plants utilizing the combined gas-steam cycle.

12.2) Main components

The key components of the parabolic trough solar power plants are primarily related to the solar field:

- Solar collectors (supporting construction);
- Tube receivers;
- Reflective panels (mirrors);
- Sun tracking system.

The remaining parts of the solar field, e.g. concrete foundation, connective piping, electrical connections etc.) are normal industrial installations, and can be qualified in an appropriate way for use in solar plants.

Other components of the heat accumulation section, such as tanks, circulating pumps and solar steam generator are custom-made for each design.

The parabolic-shaped mirrors are constructed by forming a sheet of reflective material into a parabolic shape that concentrates incoming sunlight onto a central receiver tube at the focal line of the collector.

The arrays of mirrors can be 100 metres (m) long or more, with the curved aperture of 5 m to 6 m. A single-axis tracking mechanism is used to orient both solar collectors and heat receivers toward the sun.

PTC are usually aligned North-South and track the sun as it moves from East to West to maximise the collection of energy.

The receiver comprises the absorber tube (usually metal) inside an evacuated glass envelope.

The absorber tube is generally a coated stainless steel tube, with a spectrally selective coating that absorbs the solar (short wave) irradiation well, but emits very little infrared (long wave) radiation.

This helps to reduce heat loss. Evacuated glass tubes are used because they help to reduce heat losses.

A heat transfer fluid (HTF) is circulated through the absorber tubes to collect the solar energy and transfer it to the steam generator or to the heat storage system, if any. Most existing parabolic troughs use synthetic oils as the heat transfer fluid, which are stable up to 400°C. New plants under demonstration use molten salt at 540°C either for heat transfer and/or as the thermal storage medium.

High temperature molten salt may considerably improve the thermal storage performance.

12.3) Collectors

The solar collector structure includes a part anchored to the soil (pylons) and a mobile part, supporting the reflective panels and receiving tube (Figure 5.44).

Basic characteristics of solar collector are: geometrical precision, size rigidity, resistance to bad weather, especially wind gusts and temperature fluctuations, reduced material consumption and simplicity of mounting at the construction site.

Main commercial collectors are: Luz system, EuroTrough, Solargenix, Luz System Collectors. In addition, there is a model developed by ENEA in an array of versions, sold by many companies, including Ronda, Reflex and DD.

The Luz collector is the basic system from which all others were derived. It is made of zinc-coated steel, and was used in the major part of SEGS (first generation solar plants). It is produced in two versions: LS-2 and LS-3.

LS-2 version has a structure with torsion tubes, easy to mount and with considerable torsional rigidity. It consists of six modules, three of which are on each side of the Sun tracking engine; each module hosts two serially mounted receiving tubes, each 4 meters in length.

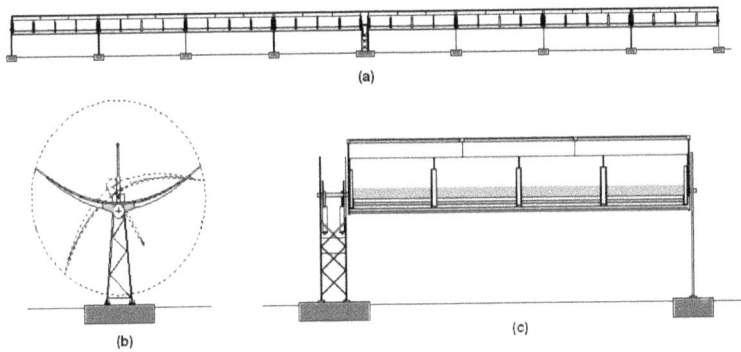

Figure 12.m – Solar collector: (a) longitudinal view, (b) lateral view, (c) enlarged part of the longitudinal view

The structure of the torsion tube has a high manufacturing price, because, in order to obtain its functional efficiency, it uses a high amount of material and requires higher manufacturing precision.

In an attempt to reduce the manufacturing costs, the Luz company has later developed the LS-3 model with a reticular structure, requiring less tolerance in manufacturing precision and using less material; however, the cost reduction turned out to be lower than expected, while lower torsion rigidity had a negative influence on optical and thermal performances.

A different solution, developed by European consortium EuroTrough, is based on the box-shaped structure that can guarantee high torsion rigidity with lower material consumption.

The Solargenix collector, produced by the US Department of Energy and NRel, is made of extruded aluminum, using a special joint system, initially developed for buildings and bridges, is extremely light and makes mounting operation simpler.

The model developed by ENEA has a torsion tube structure, whereby the manufacturing technology is optimized using processing with numerical control; its performances are provided in the Table 12.n.

Table 12.n – Main characteristics of the ENEA concentrator

Concentration efficiency[1]	> 80%
Tracking error[2]	< 7 mrad
Maximal wind velocity	37 m/s
Expected technical life	25 years

[1] Diameter 70 mm
[2] With a 7 m/s wind

12.4) Receiving tube

The receiving tube, placed on the focal line of the parabolic trough collector has functions to absorb the concentrated solar energy and to transfer it in the form of high temperature thermal energy onto the working fluid running through it.

This apparently simple function is actually very complex: if a normal tube was used, a major part of the energy would be lost through the heat exchange due to the sizeable difference in temperatures of the tube surface and the outer air, reducing its efficiency to unacceptable levels.

Therefore it is necessary to use a special component requiring special manufacturing technologies.

A receiving tube consists of the inner steel tube closed into a coaxial glass tube; the inter-space between two tubes is in vacuum, with a special air- tight glass-metal junction, and compensators of differential dilatation between glass and steel (Figure 5.45).

The outer steel surface is treated with a special 'spectrally selective' coating, i.e. allowing a high absorption of solar irradiance and low emission of infrared radiation emitted by high-temperature steel.

The outer surface of the glass is treated with an antireflective coating, maximizing the solar irradiance flow through the glass.

To eliminate the smallest amounts of gas in the interspace, special materials are utilized, the so-called getters, which absorb hydrogen and other gases that are released in traces from the metal during operation at high temperatures.

The first receiving tube developed by Luz was not enough reliable in the part of the glass-metal joint with respect to vacuum sealing. Later, Solel (ex Luz) and Schott developed new models with fundamental improvements with respect to reliability and performances.

The ENEA agency developed its own receiving tube model with advanced performances, commercialized by Archimede Solar Energy, under the ENEA license. The exclusive characteristic of such receiving tube is its capacity to operate on temperatures up to 550 °C.

The efficiency at the collector loop 600 meters in length, with a direct solar irradiance of 900 W/m² and working fluid with temperatures between 290 and 550 °C, reaches 90.8% compared to the concentrated solar energy.

Figure 12.0 – Receiving tube diagram

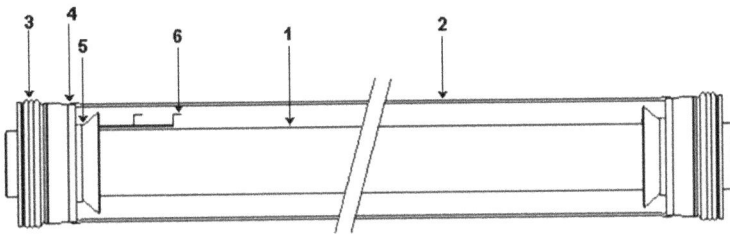

1 – Steel tube with a selective coating
2 – Outer glass tube
3 – Differential dilation compensator between glass and steel
4 – Glass-steel joint
5 – Shielding
6 – Gas absorber (Getter)

12.5) Reflective panels

Reflective panels are curved mirrors with a parabolic profile that have the function to concentrate direct solar irradiance and reflect it onto the receiving tube surface.

They consist of special glass with high transmittance (exceeding 93%), with the back surface coated by silver and multi-layer cover on the back side to protect the silver coating.

Each panel usually covers an area of around 2 m².

The main solutions include: a) thick glass (around 4 mm) hot-curved and anchored directly to the collector; b) thin glass (less than 1 mm) cold-curved and fixed onto the supporting panel. A third solution, currently under research, utilizes polished aluminum surfaces protected with a transparent film.

Basic properties of reflective panels are maintaining the reflectivity over time and resistance to breaking.

The former property significantly influences the plant's performances, while the latter directly affects operating costs, due to high costs of replacement.

12.6) Solar tracking system

Solar tracking systems function to continuously rotate the solar collectors so that the axis of the parabola, that constitutes its profile, is constantly parallel to the direction of the solar irradiance. In this manner, the mirrors reflect the direct radiation exactly onto the focal line, where the receiving tube is mounted.

The precision of the direction is extremely important since even minimal movement may place the receiving tube 'out of focus', causing rapid decrease of the temperature and reduced functionality of the plant.

A solar tracking system has to guarantee a high precision of direction, it needs to have powerful engines to confront wind gusts in operating conditions and sufficient speed to quickly place the collector into a safe position in case of strong wind.

The system consists of a 'logical' part, determining the Sun's position at any moment, and a series of actuators (engines) for rotating the mobile part of the collector.

Determining the Sun's position may be conducted by computational programs that take into account the geographical position of the site, hour of the day and period of the year, using tables and a calculating algorithm, or using measuring instruments.

In general, the Sun's position is first determined by approximate calculations, and then the fine adjustments are made using instruments, regardless of the fact that the calculating programs have reached a high level of precision, usually making the adjustments unnecessary.

The solar tracking system actuator may consist of a conventional electric motor with a gear reduction (same as the LS-2 collector) or a system with hydraulic pistons (LS-3, EuroTrough Solargenix SGX-1).

Each of the collector groups, interconnected into a rigid set, has its own engine which is normally placed into the center.

12.7) Design elements

From a design perspective, concentrated solar power plants, especially thermodynamic ones, can be compared to large industrial plants such as thermal power stations; nevertheless particular attention is needed on issues related to the changeability of the Sun as a source of energy and their logical implications to operation and to stress of the material.

This is particularly so in the case of plants operating at higher temperatures, such as molten salt plants, that require attention to the issues related to the risk of solidifying and interactions with certain materials.

Specific elements for designing concentrated solar power plants depend on the adopted technology, but are generally related to the following aspects:

- Characterization of the type of service or customer.
- Site of the plant.
- Size of the solar field.
- Heat accumulation system.
- Possible integration system.

It is possible to determine peak power and nameplate power of the plant, as is schematically shown by the Figure 12.p, on the basis of the load curve, i.e. the power required by the plant, and on the basis of the solar irradiance at the site, where the plant is planned to be built, applying adequate calculation models.

Peak power allows measuring the solar field, while the nameplate power serves to measure the steam generator and power generator.

The width of the solar field is determined by the collecting area required to obtain the peak power of the plant, enhanced by the size of the inactive parts of the collector, such as brackets, tubes etc., by the maneuvering space around the collector and by the space required to avoid casting shadow of collectors over each other.

Approximately, solar field width is twice as big as the value of the collecting area.

Heat accumulation capacity, proportional to the area A in the graph, in this example corresponding to the period of 2.5 hours, is the minimal value that could be increased based on economic considerations, by reducing the energy of integration (represented as the C area in the diagram).

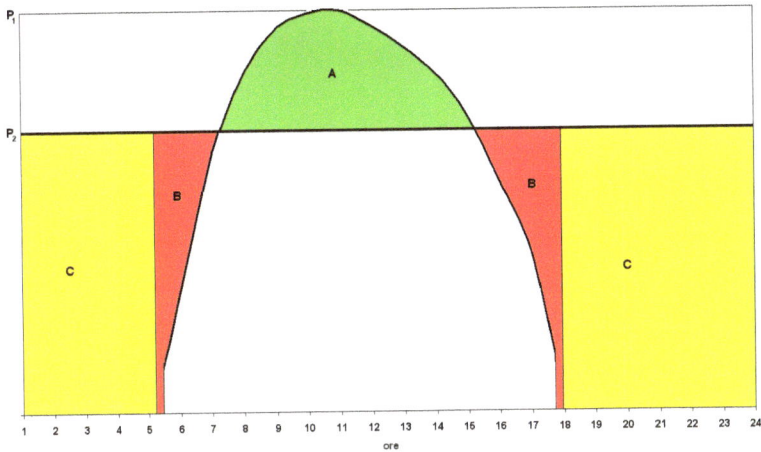

Figure 12.p – Example of an hourly diagram of the power and energy. A: Energy diverted into accumulation. B: Energy taken from accumulation. C: Integration. P_1: Peak power. P_2: Nameplate power

12.8) Industrial scale-up of technology

After a long experimental phase at the beginning of the 20th century and after the first groups of industrial scale installations in the period between 1980-1990, the practical development of the CSP plants has seen a period of stagnation until the first years of the 21st century, when interest recovered, led by the USA, Spain, Israel - and more recently expanding to many others.

The countries where the CSP technologies achieved the highest level of practical development are the USA and Spain.

In the USA, the practical development of the CSP technologies started in California at the end of 1980's, when new power plants were built with parabolic trough collectors, branded SEGS (Solar Energy Generating System), with electric power totaling around 350 MW, and plants with towers Solar One and Solar Two, with a total power of 10 MW.

After many years of stagnation, from the beginning of the 21st century, a large-scale plan to build 30 new power plants started, bringing over 8 GW of installed capacity.

From 1998, Spain started three large-scale projects: Andasol, Solar Tres and PS10.

The AndaSol 1 power plant, followed by AndaSol 2, 3 and 4, with parabolic trough collectors installed in the northern part of Sierra Nevada (Granada) have individual power of 50 MW; PS10 and subsequently PS2, both with a tower, with power 10 and 20 MW respectively, were installed near Solucar la Mayor (Seville).

The Solar Tres plant, with a tower, of 15 MW power was built near Écija, in Andalusia.

These countries were accompanied by various countries of North Africa and Middle East, including Egypt, Algeria and UA Emirates, with a total power on a global level amounting around 1064 MW.

Table 5.8 shows the major CSP plants that were in operation, under construction or at the design stage in 2010.

Table 12.q – Main concentrated solar power plants – Situation in December 2010

In operation			Under construction			In the designing stage		
Plant	Country	MW	Plant	Country	MW	Plant	Country	MW
Maricopa	Arizona	1,5	ISCC Argelia	Algeria	25	Solana	Arizona	280
Saguaro	Arizona	1	ISEGS	California	400	Abengoa Mojave	California	250
Kimberlina	California	5	Al Kuraymat	Egipat	25	Alpine SunTower	California	92
SEGS I	California	14	Alvarado I	Spain	50	Beacon	California	250
SEGS II	California	30	Andasol-3	Spain	50	Blythe	California	1000
SEGS III	California	30	Andasol-4	Spain	50	BrightSource PG&E 5	California	200
SEGS IV	California	30	Arcosol 50	Spain	50	BrightSource PG&E 6	California	200
SEGS IX	California	80	EL REBOSO II	Spain	50	BrightSource PG&E 7	California	200
SEGS V	California	30	Extresol-2	Spain	50	Calico	California	650
SEGS VI	California	30	Gemasolar	Spain	17	Gaskell	California	250
SEGS VII	California	30	Helios I	Spain	50	Genesis	California	250
SEGS VIII	California	80	Helios II	Spain	50	Imperial Valley-Solar 2	California	750
Sierra SunTower	California	5	La Dehesa	Spain	50	Palen	California	500

Cameo	Colorado	2	Lebrija 1	Spain	50	Rice	California	150
MNGSEC	Florida	75	Manchasol-1	Spain	50	Ridgecrest	California	250
Holaniku	Hawaii	2	Palma del Rio I	Spain	50	Coyote Springs 1	Nevada	200
Archimede	Italy	5	Vallesol 50	Spain	50	Coyote Springs 2	Nevada	200
ISCC Morocco	Morocco	6	Shams 1	UA Emirates	100	Sonoran	Nevada	375
Nevada Solar One	Nevada	75				Tonopah	Nevada	110
Andasol-1	Spain	50				N. M. SunTower	New Mexiko	92
Andasol-2	Spain	50				EL REBOSO III	Spain	50
Extresol-1	Spain	50				Extresol-3	Spain	50
La Florida	Spain	50				Manchasol-2	Spain	50
Majadas I	Spain	50						
Palma del Rio II	Spain	50						
PS10	Spain	11						
PS20	Spain	20						
Puerto Errado 1	Spain	1.5						
Puertollano	Spain	50						
Solnova 1	Spain	50						
Solnova 3	Spain	50						
Solnova 4	Spain	50						

In 2011, new CSP plants started operation, with total power around 545 MW, of which 20 MW in Egypt (Kuraymat), another 20 MW in Morocco (Ain Beni Mathar) and 25 MW in Algeria (Hassi R'mel), bringing thus the total power on the global level to 1,655 MW.

Other countries participating in the applicational development of CSP systems include Iran with 17 MW (Yazd), Tailand with 5 MW (Huaykrachao), Australia with 2 MW (New South Wales), Germany with 1,5 MW (Julich) and Italy with 5 MW (Priolo).

During the first months of 2012, two new 50 MW plants started operation in Spain (Solacor I and II) and a 30 MW plant (Puerto Errado II).

New plants exceeding 3 GW are planned for 2015, of which around 760 MW in the USA, 1100 MW in Spain, 250 MW in Israel (Ashalim, Negev desert), 100 MW in the United Arab Emirates (Abu Dhabi), 500 MW in Morocco (Ouarzazate), 100 MW in South Africa (Eskom), 316 MW in Australia and 28 MW in China.

The expanding trend of solar thermal systems (Figure 12.r) suggests the possibility of medium to long term spreading of these technologies in numerous countries with high insolation and marginal areas, e.g. deserts, at disposal.

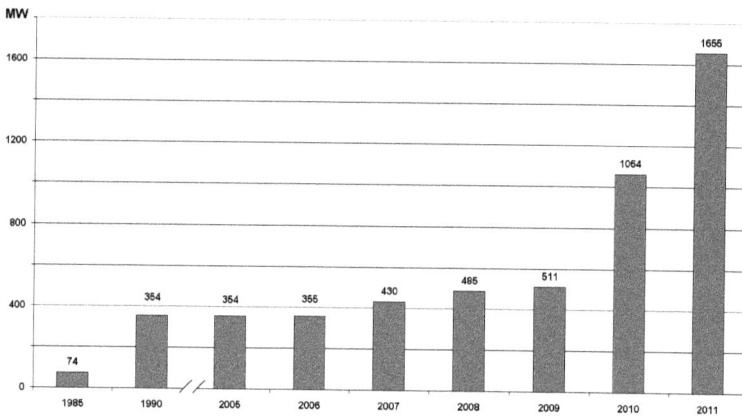

Figure 12.r – CSP plants: globally installed cumulative power

The value of solar irradiance required for the operation of a CSP plant amounts 1800-2000 kWh/m²/year in terms of DNI (direct normal irradiance), equivalent to 4.9–5.5 kWh/m²/day, which are typical values for Mediterranean climate.

The following Table 12.s shows main characteristics of CSP plant sites.

Table 12.s – Main properties of CSP plant sites

Solar irradiance	> 1800 kWh/m²/year
Area	2-3 ha/MW
Inclination	Optimal< 1% - Maximal 3%
Infrastructures	Electric lines – transport roads
Water	> 300 liter/MWh

The theoretical potential is more than enough to meet the needs for electricity of the entire planet: global electric energy consumption (around 19,000 TWh in 2009) could be provided, in general, by CSP plants covering a total area of 135.000 km², equivalent to around 1.5% of Sahara).

In theory, a great opportunity was presented by the Desertec project: in 2008, Desertec Industrial Initiative (DII), joint venture of 12 important companies, including ABB, Abengoa, Siemens, Deutsche Bank, who propose building plants with around 100 GW power using CSP technology in North Africa by 2050 and importing that power into Europe via submarine connections, covering 15% of the continent's demand, with planned investments amounting 400 billion euros.

The World Bank's Clean Technology Fund has in its program to invest 750 million dollars to build CSP plants in North Africa and in the Middle East of the power totalling 1 GW, while the Mediterranean Solar Plan of the Mediterranean Union aims to promote building CSP plants of the total power of 19 GW by 2020.

Building CSP plants in the areas of North Africa and Middle East implies the necessity to build local infrastructure and services, contributing thus to the economic and social development of these countries.

Still, some geopolitical aspects remain critical and it is necessary to solve some technical problems regarding availability of water and reliability of the system in extreme climate conditions.

As for building power plants, concentrated solar energy is a market 'by order', very similar to building large plants with a strong sector specialization, having, by its nature, global market.

Table 5.10 provides some of the major international operators.

The largest global operators in production, designing and installing are from Spain: Abengoa Solar and Acciona account for more than 50% of the total installation in 2011.

Numerous international organizations are engaged in the research of the CSP technologies (Table 12.t).

Table 12.t - Mapping of main industrial players in the CSP sector

Plant		Components	
Abengoa Solar	Spain	Archimede Solar Energy	Italy
Acciona	Spain	DD	Italy
ACS Cobra	Spain	Flabeg Int. GmbH	Germany
Black & Veatch	USA	Reflex	Italy
Bright Source Energy	USA	Rioglass	Spain
ENEL Green Power	Italy	Ronda	Italy
ENEL Ingegneria e Innovazione	Italy	Siemens	Germany
EPURON	Germany	Schott Rohrglas GmbH	Germany
E-Solar	USA	Solar Power Group	Germany
Fichtner	Germany	SolarGenix	USA
Flagsol	Germany		
FPL Energy	USA		
Iberdrola	Spain		
KT Technimont	Italy		
Lauren	USA		
Nexant	USA		
SBP Schlaich Bergermann und Partner	Germany		
Sener	Spain		
Solel Solar Systems Ltd	Israel		
South California Edison	USA		
Techint	Italy		

Main lines of research are related to enhancing the performance characteristics, finding more efficient solutions and integration with other plant types.

A special aspect regarding the CSP plant development is related to the infrastructure for transmitting the electric energy from North Africa towards Europe (Consortium Desertec).

Table 12.u – Main CSP research station

CIEMAT	Centro de Investigaciones Energéticas, Medioambientales y Tecnológicas	Spain
CNRS PROMES	Centre National de la Recherche Scientifique	France
DLR	Centro Aerospaziale Tedesco	Germany
ENEA	National Agency for New Technologies, Energy and Sustainable Economic Development	Italy
Fhg-ISE	Fraunhofer-Institut für Solare Energiesysteme	Germany
NREL	National Energy Renewable Laboratory	USA
PSA	Plataforma Solare de Almeria	Spain
PSI	Paul Scherer Institut	Switzerland
Sandia	Sandia National Laboratories	USA
WIS	Weizmann Institute of Science	Israel

12.9) Development in advanced market: Italy's example

In Italy, areas with characteristics suitable for economically viable installation of concentrated solar plants are limited to central and south regions, especially Sicily, south Apulia and a part of Sardinia.

The very possibility of using land for alternative purposes sometimes brings its availability into question.

The application potential in Italy is, thus, limited compared to Spain and countries of North Africa, but the interest for new installations is large, because it would provide an opportunity for the Italian industry to qualify for the promising international market, focused primarily to the nearby countries of North Africa and Middle East.

From the viewpoint of technological developments, Italy has a long tradition regarding the efforts in CSP, starting from the pioneering research by Prof. Francia, as early as from 1960-1970.

In1980, an Italian-French-German consortium has built in Adrano, Sicily, an experimental 1 MW solar power plant Eurelios.

In 2001, a thermodynamic solar project by ENEA (National Agency for New Technologies, Energy and Sustainable Economic Development), which led to the development of an innovative version of parabolic trough collector technology and its demonstration on industrial scale-up basis (5 MW), in the power plant of the ENEL company with a combined cycle power plant in Priolo Gargallo (the Archimede Project).

ENEA's projects in solar energy have brought about the development of the Italian industry of components for CSP plants and qualification of the ENEL Group as a constructor and operator for such plants.

Currently, there are various initiatives for building new plants including projects ARCHETYPE SW 550 and ARCHETYPE 30+ presented by ENEL Green Power with ENEA and Italian companies within the initiative NER 300 [5], participation of the ENEL Green Power company in the bid for the 'Ouarzazat' project in Morocco and MATS project by ENEA in Egypt.

12.10) New Applications

The main application of the CSP technology, especially solar thermodynamics, is in large-scale electricity-generating plants, preferably located in deserts.

In this type of application, thermodynamic solar power shows its potential to the fullest, especially reducing unit costs as a result of economies of scale, but also regularity and programmability of production.

---

[5] NER 300 is a financing instrument managed jointly by European Commission, European Investment Bank and Member States. It is based on the agreement of the European Union regarding the GHG reduction through obligation for the companies using fossil fuels to obtain permit to emission; the income from the sale of 300 million permits (each one corresponding to a ton of CO_2), equivalent to around 2.4 billion euros, would be used to subsidize installing innovative technologies for renewable energies and for carbon capture and storage (CCS). Project ARCHETYPE SW 550 plans building a CSP plant integrated with a desalination plant and a 30 MW solar plant ARCHETYPE 30+, able to provide thermal energy for household use.

The development of these applications requires long periods of time, especially in the initial phase, and their notable economic size limits the market to large companies.

Stimulated by smaller-size companies, the development of adequate solutions for small and medium size plants has begun recently, and is likely to reach niches of economic benefit in conditions different from those required by large companies.

Strategies for reaching economic sustainability depend on reducing plant costs by simplification and standardization of the system on one hand, and on appreciating production through using thermal energy from waste.

In this category, plants at the developmental stage are systems called mini CSP or micro CSP: they are thermodynamic solar plants with up to 10 MW and 1 MW, respectively.

The technology utilized is mostly parabolic trough collector technology with small-size mirrors, able to reach temperatures lower than the larger plants.

The application is focused to both commercial and industrial sectors, with plants that may be installed even on the roof covers of industrial or commercial buildings for cogeneration (generating electrical and thermal energy), for industrial processing or air conditioning.

Sopogy, a US company, started operations in this sector in 2002, and more recently, certain Italian companies such as D&D from Udine, FERA from Milan and Xeliox, from the Donati group, are engaged in developing specific products for these applications.

An example in point is a 2 MW plant built in Hawaii at Natural Energy Laboratory from Keahole Point by Keahole Solar Power with Sopogy collectors, able to reach 176°C.

12.11) Economic aspects

CSP technologies can at least partially replace use of fossil fuels in generation of electric energy, hence the mid- to long term target is reaching comparable production cost, so-called grid parity.

Since the CSP plants do not consume fuel, their operating and maintenance costs have a relatively limited share; economic benefits are connected primarily with the value of the funds required for their manufacture.

Plant costs per unit electric power installed (€/MW) depend on many factors, primarily on the plant size and the capacity of the heat accumulation system: the former as a consequence of the strong effect of the scale to the conventional part of the plant (thermal cycle, plant installations etc.) and, the latter, apart from additional expenses for tanks and thermal fluid for accumulation, primarily is a result of oversized solar field compared to how much it requires to generate the nameplate power.

As CSP plants become more widespread, installation costs decrease to a significant extent. For example, as for to 2011, it is possible to establish a 50 MW plant with an 8 hour accumulation with funds of order of magnitude 5 € per MW.

The cost of electric energy produced by CSP plants depends on the initial investment, annual production and operation and maintenance cost annually.

There is a tendency to reduce the plant manufacturing costs and to increase productivity through technological improvements.

As for 2010, the levelized cost of electricity (LCOE) is determined between 17 and 24 c€/kWh, compared to 6-8 c€/kWh for plants powered by fossil fuels, i.e. between double and quadruple, depending on the adopted technology and location of the plant. With reduced plant costs, increase in efficiency and economies of scale, it is foreseen that they will be at the same level by 2025.

Currently, the economic balance of these plants requires public financing or other forms of incentives; its benefits are compensated a great deal by advantages related to diversification of the energy source provided, environmental concerns and development of local economies.

Among the economic benefits, related to building, such plants it should be considered that, compared to other innovative technologies, the component of conventional materials and work (cement works, metal works etc.) accounts for a considerable part of the overall investment, which enhances local economies.

CSP technologies provide an opportunity for both investors and the developers of the technology: wide availability of the solar source in many countries offers scope to join a global market of a notable size.

Medium to large size plants may seem difficult to find funding in times of crisis, but a sufficient turnover can be guaranteed to cover the investments into research and development.

Compared to photovoltaic plants, the difference in the anticipated costs is very high in the beginning, in favor of CSP, and then it decreases progressively as a result of powerful commercial development of photovoltaic plants.

It can be considered that the two technologies will have complementary roles: photovoltaic will be in a wide use, owing to the simplicity of the plant and the fact that it does not require special space (installations on roof covers and walls), while the CSP technology will be used in large-scale concentrated installations, utilizing economies of scale the most and the advantage of thermal accumulation (increased number of production hours compared to availability of solar irradiance, effective power close to the installed power, regularity and programmability of power generation).

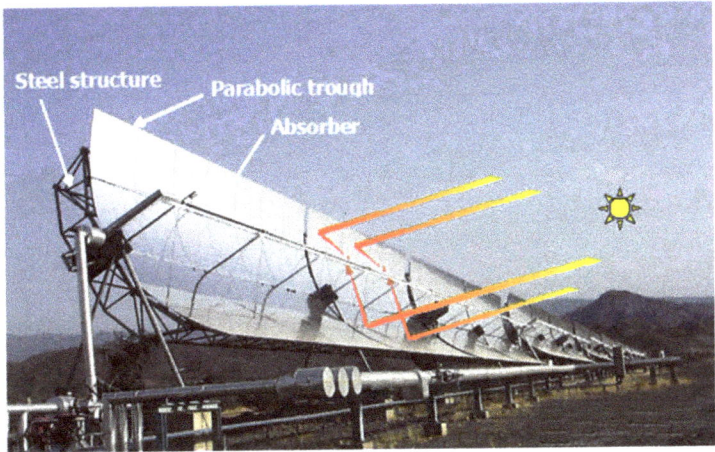

Picture 12.v - Parabolic trough collector technology

12.12) Linear Fresnel collector

Linear Fresnel collectors (LFCs) are similar to parabolic trough collectors, but use a series of long flat, or slightly curved, mirrors placed at different angles to concentrate the sunlight on either side of a fixed receiver (located several meters above the primary mirror field).

Each line of mirrors is equipped with a single-axis tracking system and is optimized individually to ensure that sunlight is always concentrated on the fixed receiver.

The receiver consists of a long, selectively-coated absorber tube.

Unlike parabolic trough collectors, the focal line of Fresnel collectors is distorted by astigmatism.

This requires a mirror above the tube (a secondary reflector) to refocus the rays missing the tube, or several parallel tubes forming a multi-tube receiver that is wide enough to capture most of the focused sunlight without a secondary reflector.

The main advantages of linear Fresnel CSP systems compared to parabolic trough systems are that:

> LFCs can use cheaper flat glass mirrors, which are a standard mass-produced commodity;

> LFCs require less steel and concrete, as the metal support structure is lighter. This also makes the assembly process easier;

> The wind loads on LFCs are smaller, resulting in better structural stability, reduced optical losses and less mirror-glass breakage and

> The mirror surface per receiver is higher in LFCs than in PTCs, which is important, given that the receiver is the most expensive component in both PTC and in LFCs.

These advantages need to be balanced against the fact that the optical efficiency of LFC solar fields (referring to direct solar irradiation on the cumulated mirror aperture) is lower than that of PTC solar fields due to the geometric properties of LFCs.

The problem is that the receiver is fixed and in the morning and afternoon cosine losses are high compared to PTC.

Despite these drawbacks, the relative simplicity of the LFC system means that it may be cheaper to manufacture and install than PTC CSP plants. However, it remains to be seen if costs per kWh are lower.

Additionally, given that LFCs are generally proposed to use direct steam generation, adding thermal energy storage is likely to be more expensive.

Picture 12.z - Fresnel linear reflector

12.13) Solar Tower Technology

Solar tower technologies use a ground-based field of mirrors to focus direct solar irradiation onto a receiver mounted high on a central tower where the light is captured and converted into heat.

The heat drives a thermo-dynamic cycle, in most cases a water-steam cycle, to generate electric power.

The solar field consists of a large number of computer-controlled mirrors, called heliostats, that track the sun individually in two axes.

These mirrors reflect the sunlight onto the central receiver, where a fluid is heated up. Solar towers can achieve higher temperatures than parabolic trough and linear Fresnel systems, because more sunlight can be concentrated on a single receiver and the heat losses at that point can be minimized.

Current solar towers use water/steam, air or molten salt to transport the heat to the heat-exchanger/steam turbine system.

Depending on the receiver design and the working fluid, the upper working temperatures can range from 250°C to perhaps as high 1.000°C for future plants, although temperatures of around 600°C will be the norm with current molten salt designs.

The typical size of today's solar tower plants ranges from 10 MW to 50 MW. The solar field size required increases with annual electricity generation desired, which leads to a greater distance between the receiver and the outer mirrors of the solar field.

This results in increasing optical losses due to atmospheric absorption, unavoidable angular mirror deviation due to imperfections in the mirrors and slight errors in mirror tracking.

Solar towers can use synthetic oils or molten salt as the heat transfer fluid and the storage medium for the thermal energy storage. Synthetic oils limit the operating temperature to around 390°C, limiting the efficiency of the steam cycle.

Molten salt raises the potential operating temperature to between 550 and 650°C, enough to allow higher efficiency supercritical steam cycles, although the higher investment costs for these steam turbines may be a constraint.

An alternative is direct steam generation (DSG), which eliminates the need and cost of heat transfer fluids, but this is at an early stage of development and storage concepts for use with DSG still need to be demonstrated and perfected.

Solar towers have a number of potential advantages, which mean that they could soon become the preferred CSP technology.

The main advantages are that:

> The higher temperatures can potentially allow greater efficiency of the steam cycle and reduce water consumption for cooling the condenser;

> The higher temperature also makes the use of thermal energy storage more attractive in order to achieve schedulable power generation and

> Higher temperatures will also allow greater temperature differentials in the storage system, reducing costs or allowing greater storage for the same cost.

The key advantage is the opportunity to use thermal energy storage to raise capacity factors and allow a flexible generation strategy to maximise the value of the electricity generated, as well as to achieve higher efficiency levels.

Given this advantage and others, if costs can be reduced and operating experience gained, solar towers could potentially achieve significant market share in the future, despite PTC systems having dominated the market to date.

Solar tower technology is still under demonstration, with 50 MW scale plant in operation, but could in the long-run provide cheaper electricity than trough and dish systems. However, the lack of commercial experience means that this is by no means certain and deploying solar towers today includes significant technical and financial risks.

Picture12.w - Solar Tower Technology

12.14 Thermal energy storage

Thermal energy storage (TES) is achieved with greatly differing technologies that collectively accommodate a wide range of needs.

It allows excess thermal energy to be collected for later use, hours, days or many months later, at individual building, multiuser building, district, town or even regional scale depending on the specific technology.

As examples: energy demand can be balanced between day time and night time; summer heat from solar collectors can be stored inter-seasonally for use in winter; cold one, obtained from winter air can be provided for summer air conditioning. Storage mediums include: water or ice-slush tanks ranging from small to massive, masses of native earth or bedrock accessed with heat exchangers in clusters of small-diameter boreholes (sometimes quite deep); deep aquifers contained between impermeable strata; shallow, lined pits filled with gravel and water and top-insulated; and eutectic, phase-change materials.

Other sources of thermal energy for storage include heat or cold produced with heat pumps from off-peak, lower cost electric power, a practice called peak shaving; heat from combined heat and power (CHP) power plants; heat produced by renewable electrical energy that exceeds grid demand and waste heat from industrial processes.

Most practical active solar heating systems have storage for a few hours to a day's worth of energy collected.

There is a growing number of facilities that use seasonal thermal energy storage (STES), enabling solar energy to be stored in summer (primarily) for space heating use during winter.

The Drake Landing Solar Community in Alberta, Canada has now achieved a year-round 97% solar heating fraction, a world record and possible only by incorporating STES.

Molten salt is now in use as a means to retain a high temperature thermal store, in conjunction with concentrated solar power for later use in electricity generation, to allow solar power to provide electricity on a continuous basis, as base load energy. These molten salts (Potassium nitrate, Calcium nitrate, Sodium nitrate, Lithium nitrate, etc.) have the property to absorb and store the heat energy that is released to the water, to transfer energy when needed.

To improve the salt properties it must be mixed in a eutectic mixture.

High peak loads drive the capital expenditures of the electricity generation industry. The industry meets these peak loads with low-efficiency peaking power plants, usually gas turbines, which have lower capital costs and, since the recent drop in natural gas prices have low fuel costs as well.

A kilowatt-hour of electricity consumed at night can be produced at much lower marginal cost. Utilities have begun to pass these lower costs to consumers, in the form of Time of Use (TOU) rates, or Real Time Pricing (RTP) Rates.

Stored solar thermal energy has the potential to provide cheaper peak-demand power than any other energy source.

Whatever technology is used, a 100 MW solar plant with thermal storage requires:

+ 4 km² of land
+ 25 000 tons of steel
+ 12 000 tons of glass
+ 30 000 tons of storage medium
+ 20 000 m³ of concrete

This requires transport by 4000 20t trucks or 2000 railway wagons.

Solar electricity cost of concentrating solar power plants with conventional electricity generation:

+ 3-6 €c/kWh middle load
+ 5-8 €c/kWh peak load
+ 6-9 €c/kWh niche markets with high fuel costs

Picture 12.k - Molten salt

Picture 12.x - District heating accumulation tower from Theiss near Krems an der Donau in Lower Austria with thermal capacity of 2 Gwh

Molten salt can be employed as a thermal energy storage method to retain thermal energy collected by a solar tower or solar trough, so that it can be used to generate electricity in bad weather or at night.

It was demonstrated in the Solar Two project from 1995-1999.

Picture 12.j - Solarthermal power plant

The system is predicted to have an annual efficiency of 99%, a reference to the energy retained by storing heat before turning it into electricity, versus converting heat directly into electricity. The molten salt mixtures vary.

The most extended mixture contains sodium nitrate, potassium nitrate and calcium nitrate.

It is non-flammable and nontoxic, and has already been used in the chemical and metals industries as a heat-transport fluid, so experience with such systems exists in non-solar applications.

The salt melts at 131 °C (268 °F). It is kept liquid at 288 °C (550 °F) in an insulated "cold" storage tank.

The liquid salt is pumped through panels in a solar collector where the focused sun heats it to 566 °C (1,051 °F).

It is then sent to a hot storage tank. This is so well insulated that the thermal energy can be usefully stored for up to a week.

When electricity is needed, the hot salt is pumped to a conventional steam-generator to produce superheated steam for a turbine/generator as used in any conventional coal, oil or nuclear power plant.

A 100 MW turbine would need a tank of about 30 feet (9.1 m) tall and 80 feet (24 m) in diameter to drive it for four hours by this design.

Several parabolic trough power plants in Spain and solar power tower developer SolarReserve use this thermal energy storage concept.

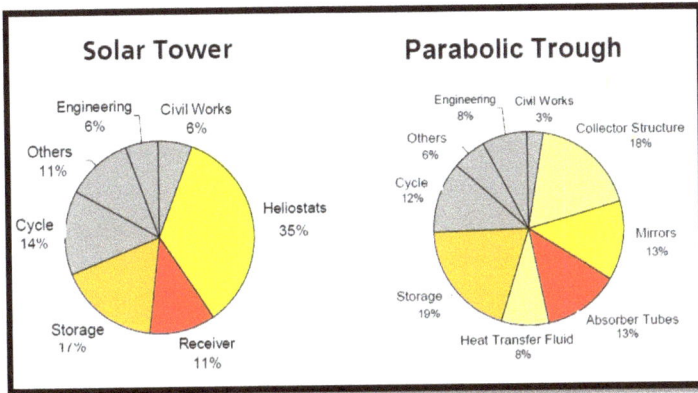

Figure 12.α – Cost structure of Solarthermal Power Plants

13.15) Stirling dish technology

The Stirling dish system consists of a parabolic dishaped concentrator (like a satellite dish) that reflects direct solar irradiation onto a receiver at the focal point of the dish. The receiver may be a Stirling engine (dish/engine systems) or a micro-turbine.

Stirling dish systems require the sun to be tracked in two axes, but the high energy concentration onto a single point can yield very high temperatures.

Stirling dish systems are yet to be deployed at any scale.

Most research is currently focussed on using a Stirling engine in combination with a generator unit, located at the focal point of the dish, to transform the thermal power to electricity.

There are currently two types of Stirling engines: kinematic and free piston. Kinematic engines work with hydrogen as a working fluid and have higher efficiencies than free piston engines.

Free piston engines work with helium and do not produce friction during operation, which enables a reduction in required maintenance.

The main advantages of Stirling dish CSP technologies are that:

➢ The location of the generator - typically, in the receiver of each dish - helps reduce heat losses and means that the individual dish-generating capacity is small, extremely modular (typical sizes range from 5 to 50 kW) and are suitable for distributed generation;

➢ Stirling dish technologies are capable of achieving the highest efficiency of all types of CSP systems;

➢ Stirling dishes use dry cooling and do not need large cooling systems or cooling towers, allowing CSP to provide electricity in water-constrained regions and

➢ Stirling dishes, given their small foot print and the fact they are self-contained, can be placed on slopes or uneven terrain, unlike PTC, LFC and solar towers.

These advantages mean that Stirling dish technologies could meet an economically valuable niche in many regions, even though the levelised cost of electricity is likely to be higher than other CSP technologies. Apart from costs, another challenge is that dish systems cannot easily use storage.

Stirling dish systems are still at the demonstration stage and the cost of mass-produced systems remains unclear. With their high degree of scalability and small size, stirling dish systems will be an alternative to solar photovoltaics in arid regions.

Picture 12.β - Stirling Dish Technology

12.16) *Solar updraft tower*

The solar updraft tower (SUT) is a renewable-energy power plant for generating electricity from solar power. Sunshine heats the air beneath a very wide greenhouse-like roofed collector structure surrounding the central base of a very tall chimney tower.

The resulting convection causes a hot air updraft in the tower by the chimney effect. This airflow drives wind turbines placed in the chimney updraft or around the chimney base to produce electricity.

Plans for scaled-up versions of demonstration models will allow significant power generation, and may allow development of other applications, such as water extraction or distillation, and agriculture or horticulture.

As a solar chimney power plant (SCPP) proposal for electrical power generation, commercial investment is discouraged by the high initial cost of building a very large novel structure and by the risk of investment in a feasible but unproven application of even proven component technology for long-term returns on investment—especially when compared to the proven and demonstrated greater short-term returns on lesser investment in coal-fired or nuclear power plants.

Likewise, the benefits of 'clean' or solar power technologies are shared, and the widely shared harmful pollution of existing power generation technologies is not applied as a cost for private commercial investment.

This is a well-described economic trade-off between private benefit and shared cost, versus shared benefit and private cost.

If it is in the public interest, then some form of public investment or subsidy to share cost and risk will be required to demonstrate SCPP feasibility at scale.

Power output depends primarily on two factors: collector area and chimney height.

A larger area collects and warms a greater volume of air to flow up the chimney; collector areas as large as 7 kilometres (4.3 mi) in diameter have been discussed.

A larger chimney height increases the pressure difference via the stack effect; chimneys as tall as 1,000 metres (3,281 ft) have been discussed.

Due to variations in design, climate, local geography and latitude, a standardized model for comparisons between design features and outputs is needed and proposed.

Heat can be stored inside the collector area.

The ground beneath the solar collector, water in bags or tubes, or a saltwater thermal sink in the collector could add thermal capacity and inertia to the collector.

Humidity of the updraft and condensation in the chimney could increase the energy flux of the system.

Turbines with a horizontal axis can be installed in a ring around the base of the tower, as once planned for an Australian project and seen in the diagram above; or—as in the prototype in Spain—a single vertical axis turbine can be installed inside the chimney.

Carbon dioxide is emitted only negligibly as part of operations.

Manufacturing and construction require substantial power, particularly to produce cement.

Net energy payback is estimated to be 2–3 years.

Since solar collectors occupy significant amounts of land, deserts and other low-value sites are most likely.

A small-scale solar updraft tower may be an attractive option for remote regions in developing countries.

The relatively low-tech approach could allow local resources and labour to be used for construction and maintenance.

Locating a tower at high latitudes could produce up to 85 per cent of the output of a similar plant located closer to the equator, if the collection area is sloped significantly toward the equator.

The sloped collector field is built on suitable mountainsides, which also functions as a chimney.

A short vertical chimney on the mountaintop to accommodate the vertical axis air turbine.

The results showed that solar chimney power plants at high latitudes may have satisfactory thermal performance.

Solar updraft towers can be combined with other technologies to increase output. Solar thermal collectors or photovoltaics can be arranged inside the collector greenhouse.

This could further be combined with agriculture.

Picture 12.γ - Comparing heights of buildings   *Picture 12.δ - Futuristic look of solar*

The solar updraft tower has a power conversion rate considerably lower than many other designs in the (high temperature) solar thermal group of collectors.

The low conversion rate is balanced to some extent by the lower cost per square metre of solar collection.

Model calculations estimate that a 100 MW plant would require a 1,000 m tower and a greenhouse of 20 square kilometres (7.7 sq mi). A 200 MW tower with the same tower would require a collector 7 kilometres in diameter (total area of about 38 km²).

One 200MW power station will provide enough electricity for around 200,000 typical households and will abate over 900,000 tons of greenhouse producing gases from entering the environment annually.

The collector area is expected to extract about 0.5 percent, or 5 W/m² of 1 kW/m², of the solar energy that falls upon it.

Concentrating thermal (CSP) or photovoltaic (CPV) solar power plants range between 20% to 31.25% efficiency (Stirling dish).

Overall CSP/CPV efficiency is reduced because collectors do not cover the entire footprint. Without further tests, the accuracy of these calculations is uncertain.

The performance of an updraft tower may be degraded by factors such as atmospheric winds, by drag induced by the bracings used for supporting the chimney and by reflection off the top of the greenhouse canopy.

Year	1984	1985	1989	1990	...	2006	2007	2008	2009	2010	2011	2012
Installed	14	60	200	80	0	1	74	55	178.5	306.5	628.5	802.5
Cumulative	14	74	274	354	354	355	429	484	662.5	969	1597.5	2553

Table 12.ε – Installed and Cumulative CSP over the years

13. OLED technology

An OLED or organic light-emitting diode is a light-emitting diode (LED) in which the emissive electroluminescent layer is a film of organic compound which emits light in response to an electric current.

This layer of organic semiconductor is situated between two electrodes.

Generally, at least one of these electrodes is transparent.

OLEDs are used to create digital displays in devices such as television screens, computer monitors, portable systems such as mobile phones, handheld games consoles and PDAs.

A major area of research is the development of white OLED devices for use in solid-state lighting applications.

OLED technology in the way of energy harvest are new for market. It is now used as display for mobile smart-phones.

It works in way that electricity is converted to light, but it also can work backwards, light to electricity.

OLEDs are flexible, they produce a lot of light per watt and they can be mass produced inexpensively.

When they are exposed to a bright light, the reaction makes current to flow out of the OLED instead of into it.

So they can be energy collectors and light emitters, depending on the needs of the consumer.

Team of researchers at the University of Cambridge are working on a project to harvest energy from wasted light in an OLED display.

Their idea is that OLED screen use solar cells to absorb scattered and wasted light, sending it back into the screen.

In general only 36% of light produced by OLED display is projected forwards and the rest escapes around the edges in the form of scatter and bleeding from the edges.

The researchers worked on a solution where they could harvest what's lost by installing photovoltaic cell on the back and sides of OLED screens to capture the loss.

This is the way towards the energy decentralization in means that we will not have to plug our phone to charger everyday but at least once per week !

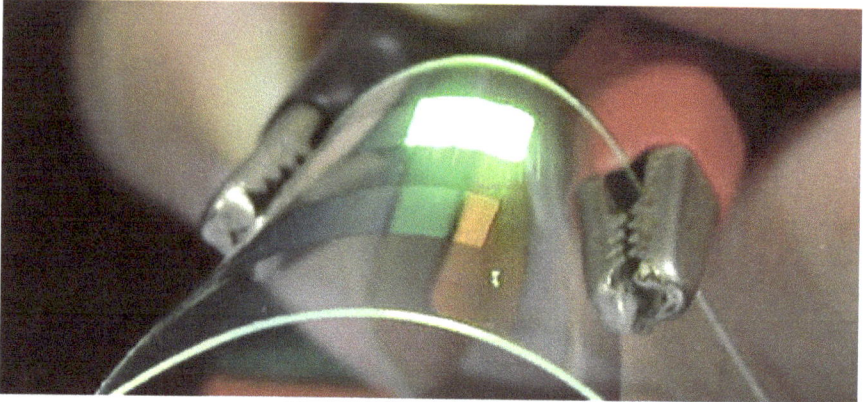

Picture 13.a - OLED flexible cell

14. Global markets and challenges of solar systems

14.1) PV, CPV and Thin film

The solar photovoltaic (PV) market saw another strong year, with total global operating capacity reaching the 100 GW milestone in 2012.

The market was fairly stable relative to 2011, with slightly less capacity brought on line but likely higher shipment levels, and the more than 29.4 GW added represented nearly one-third of total global capacity in operation at year's end.

The thin film market share fell from 15% in 2011 to 13% in 2012.

Eight countries added more than 1 GW of solar PV to their grids in 2012, and the distribution of new installations continued to broaden.

The top markets, like Germany, Italy, China, the United States and Japan, were also the leaders for total capacity.

By year's end, eight countries in Europe, three in Asia, the United States and Australia had at least 1 GW of total capacity.

The leaders for solar PV per inhabitant were Germany, Italy, Belgium, the Czech Republic, Greece and Australia.

Europe again dominated the market, adding 16.9 GW and accounting for about 57% of newly installed capacity, to end 2012 with 70 GW in operation.

But additions were down from 22 GW and more than 70% of the global market in 2011; the region's first market decline since at least 2000 was due largely to reduced incentives (including FiT scheme payments) and general policy uncertainty, with the most significant drop in Italy.

Regardless, for the second year running the EU installed more PV than any other electricity-generating technology: PV represented about 37% of all new capacity in 2012.

As its share of generation increases, PV is starting to affect the structure and management of Europe's electricity system, and is increasingly facing barriers that include direct competition with conventional electricity producers and saturation of local grids.

Italy and Germany both ended 2012 with more solar PV than wind capacity in operation, together accounting for almost half of the global total.

Germany added a record 7.6 GW, up just slightly over the previous two years, increasing its total to 32.4 GW.

SOLAR PHOTOVOLTAICS (PV)

SOLAR PV GLOBAL CAPACITY, 1995–2012

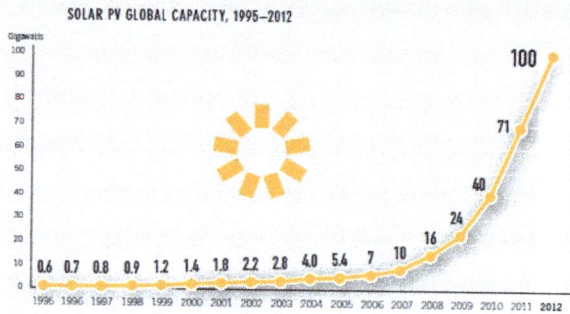

Figure 14.a - Solar PV global capacity over the years

Solar PV generated 28 TWh of electricity in Germany during 2012, up 45% over 2011 Italy reached a total capacity of 16.4 GW; however, the 3.6 GW brought on line was far lower than additions in 2011.

Other top EU markets included France (1.1 GW), the United Kingdom (0.9 GW), Greece (0.9 GW), Bulgaria (0.8 MW), and Belgium (0.6 MW).

All saw total operating capacity increase 30% or more, with Bulgaria's capacity rising six fold, although France's market was down relative to 2011.

Beyond Europe, about 12.5 GW was added worldwide, up from 8 GW in 2011.

The largest markets were China (3.5 GW), the United States (3.3 GW), Japan (1.7 GW), Australia (1 GW), and India (almost 1 GW).

Asia (7 GW) and North America (3.6 GW) followed Europe for capacity added; by year's end, Asia was rising rapidly and was second only to Europe for total operating capacity.

U.S. capacity was up nearly 85% in 2012 to 7.2 GW.

MARKET SHARES OF TOP 15 SOLAR PV MODULE MANUFACTURERS, 2012

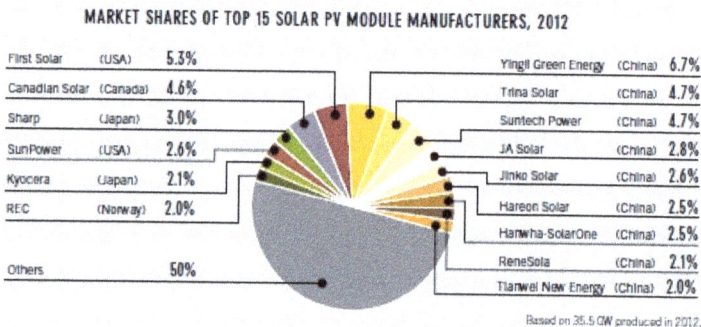

First Solar	(USA)	5.3%
Canadian Solar	(Canada)	4.6%
Sharp	(Japan)	3.0%
SunPower	(USA)	2.6%
Kyocera	(Japan)	2.1%
REC	(Norway)	2.0%
Others		50%
Yingli Green Energy	(China)	6.7%
Trina Solar	(China)	4.7%
Suntech Power	(China)	4.7%
JA Solar	(China)	2.8%
Jinko Solar	(China)	2.6%
Hareon Solar	(China)	2.5%
Hanwha-SolarOne	(China)	2.5%
ReneSola	(China)	2.1%
Tianwei New Energy	(China)	2.0%

Based on 35.5 GW produced in 2012.

Figure 14.b - Solar PV global capacity, by Countries

California had a record year (>1 GW added) and was home to 35% of total U.S. capacity. But PV is spreading to more states, driven by falling prices and innovative financing and ownership models such as solar leasing, community solar investments, and third-party financing.

On the negative side, battles are emerging around the future of net metering due to utility concerns about potential stranded costs of existing generating assets and related infrastructure.

Utility installations represented 54% of additions and accounted for 2.7 GW of U.S. capacity by year's end, with more than 3 GW under construction.

Utility procurement is slowing, however, as many utilities approach their Renewable Portfolio Standard (RPS) targets.

China doubled its capacity, ending 2012 with about 7 GW, but below expectations for the year. By the fourth quarter, China accounted for more than a third of global panel shipments, surging past Germany in response to government efforts to create a market for the glut of domestic solar panels.

The market is dominated by large-scale ground-mounted systems, many of which are in western China, far from load centers.

But national policies aim to encourage distributed, building-mounted projects as well.

By early 2013, about 90 plants in operation were larger than 30 MW and some 400 had at least 10 MW of capacity.

The world's 50 biggest plants reached cumulative capacity exceeding 4 GW by the end of 2012 and, at least, 12 countries across Europe, North America and Asia had solar PV plants over 30 MW.

More than 20 of these facilities came on line in 2012, including the world's two largest: a 250 MW thin film plant in the State of Arizona (USA) and a 214 MW plant in Gujarat (India).

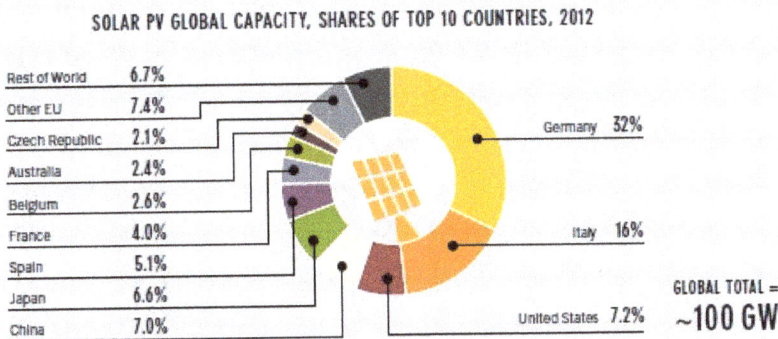

SOLAR PV GLOBAL CAPACITY, SHARES OF TOP 10 COUNTRIES, 2012

Rest of World	6.7%
Other EU	7.4%
Czech Republic	2.1%
Australia	2.4%
Belgium	2.6%
France	4.0%
Spain	5.1%
Japan	6.6%
China	7.0%

Germany 32%

Italy 16%

United States 7.2%

GLOBAL TOTAL = ~100 GW

Figure 14.c - Market share based on 35.5 GW produced in 2012

Solar PV is starting to play a substantial role in electricity generation in some countries, meeting an estimated 5.6% of national electricity demand in Italy and about 5% in Germany in 2012, with far higher shares in both countries during sunny months.

By year's end, PV capacity in the EU was enough to meet an estimated 2.6% of total consumption, and global capacity in operation was enough to produce at least 110 TWh of electricity per year.

With respect to the sector of civil and industrial building, we should underline that such a "Building-Integrated PhotoVoltaics" or B.I.P.V. consist more and more of photovoltaic materials, that are used to replace conventional building materials in parts of the building envelope, such as the roofs, windows, skylights or facades.

The most relevant upside of a large using of integrated photovoltaics, over more common non-integrated, systems is that the initial cost can be offset by reducing the money amount, spent on building materials and labour that would normally be used to construct the part of the building that the BIPV modules replace. These advantages make BIPV one of the fastest growing segments of the photovoltaic industry.

The term "Building-Applied PhotoVoltaics" or B.A.P.V. is sometimes referred to photovoltaics that are a retrofit - integrated into the building after construction is complete. Most building-integrated installations are actually BAPV.

Manufacturers and builders have to differentiate new construction BIPV from BAPV.

According to industrial report "Building-Integrated Photovoltaics Market 2012" (NanoMarkets), I like to remember the main results:

✓ Global BIPV market is set to grow by over US $5 billion by 2015.
✓ Currently worth just US$2.1 billion, the market will increase to US$7.5 billion by 2015, the industry analysts predict.
✓ The growth in BIPV market share will mostly be at the expense of conventional solar panels.
✓ 63% of total BIPV revenues are expected to come from new builds.
✓ A breakdown of the BIPV market shows that by 2015 revenues from BIPV roofing products will be worth US$2.5 billion.
✓ Revenues from the BIPV walling market will be worth US$830 million.

- ✓ Revenues generated from BIPV glass products are projected to be worth US$4.2 billion. But much of this figure will be due to the high-cost of the architectural glass that underpins BIPV glass.
- ✓ Setting aside the high prices, there appears to be good opportunities in the glass sector for monolithically integrating PV and building fabric functionalities. As a result, US$375 million will be generated by fully-integrated BIPV glass products by 2015.

From the other side, as in 2011 and 2012 was a good year for solar PV distributors, installers and consumers, we should remember that cell and module manufacturers struggled to survive let alone make a profit.

An aggressive capacity build-up in 2010 and 2011, especially in China as we saw before, resulted in excess production capacity and supply that, alongside extreme competition, drove prices down further in 2012, yielding smaller margins for manufacturers and spurring continued industry consolidation.

Low prices also have challenged many thin film companies and the concentrating solar industries, which are struggling to compete.

The average price of crystalline silicon solar modules fell by 30% or more in 2012, while thin film prices dropped about 20%.

Installed system costs are also falling, although not as quickly, and they vary greatly across locations.

From the second quarter of 2008 to the same period in 2012, German residential system costs fell from USD 7.00/Watt (W) to USD 2.20/W; by contrast, average prices for U.S. residential systems had fallen to USD 5.50/W.

Approximately 31.9 GW of crystalline silicon cells and 35.5 GW of modules were produced in 2012, down slightly from 2011.

Despite several plant closures, year-end module production capacity increased in 2012, with estimates ranging from below 60 GW to well over 70 GW.

China's production capacity alone exceeded the global market.

Thin film production declined nearly 15% in 2012, to 4.1 GW and its share of total global PV production continued to fall.

Over the past decade, leadership in module production has shifted from the United States, to Japan, to Europe, to Asia. By 2012, Asia accounted for 86% of global production (up from 82% in 2011), with China producing almost two-thirds of the world total.

Europe's share continued to fall, from 14% in 2011 to 11% in 2012, and Japan's share dropped from 6% to 5%.

The U.S. share remained at 3%; thin film accounted for 29% of U.S. production, down from 41% in 2011.Europe was still competitive for polysilicon production, however and the United States was the leading producer.

More than 24 U.S. solar manufacturers have left the industry in recent years, and, by one estimate, about 10 European and 50 Chinese manufacturers went out of business during 2012. Even "tier 1" Chinese companies like Yingli and Trina idled plants and struggled to stay afloat.

By year's end, China's 10 largest manufacturers had borrowed almost USD 20 billion from state-owned banks and Suntech Power's main operating subsidiary declared bankruptcy in early 2013.

In India, 90% of domestic manufacturing had closed or filed for debt restructuring by early 2013.

The year 2012 was also mixed for CPV. Several companies, including Skyline Solar and GreenVolts (both USA), closed their doors and SolFocus (USA) announced a decision to sell; but those companies that were still operating invested increasing amounts of time and money in building manufacturing facilities in emerging markets. The industry is currently in the commercialization phase, but several challenges remain, including obtaining financing required to scale up projects and demonstrating continuous high yield outside the laboratory.

In fact, the CPV market is focused on three major technological sectors regarding optical concentration:

1. reflective optics,
2. refractive optics,
3. hybrid reflective/refractive optics

and two sectors for concentration level:

A. low concentration PV (LCPV) with concentration factors below 100X
B. high concentration PV (HCPV) with concentration factors of 300X and more (up to 1000X).

Medium concentration with concentration factors ranging from 100 – 200X is practically marginal.

As for cells, companies are able to provide high production rates at an industrial scale. Just one HCPV cell wafer can supply up to 1.0-1.5 kW_p or more, while the existing production rate of semiconductor manufacturers ensure up to 100-200 GW_p

equivalent solar cells per annum, without any additional effort to sharply decrease the material thickness or discover groundbreaking technical solutions.

Many companies have transferred from the stage of developing module and system prototypes to production. Companies such as SolFocus, Boeing (USA) developed solutions based on Fresnel plastic lenses, generally made of methacrylate: Amonix, (USA), Opel Solar (Canada), Suntrix (China), Arimaeco (Taiwan), Sharp (Japan), Isofoton (Spain) etc...

Daido Steel (Japan) proposes modules with dome-shaped Fresnel lenses, providing strength and reducing air pollution.

Italian companies have also established strong presence in the market, transforming prototypes into commercial solutions; such companies are serious Italian players in this sector with numerous, high quality technological solutions.

Until present, in Italy, the access of CPV technology to the mechanism of feed-in tariffs in accordance with "Fourth Energy Bill" lead to installing 44 CPV systems with 12.9 MW_p installed power.

The total installed power is expected to double by the end of 2012, reaching 90 MW, with a rapid growth perspective in the next 5 years that would allow achieving around 1.2 GW by the end of 2016 [4].

USA and Central America will be the most attractive markets, along with Middle East and Africa, especially in the areas with favorable characteristics of direct solar radiation.

If by 2035 an annual production rate of 10 GW is achieved, it will be definitely possible to obtain the CPV system costs of approximately 1 $€/W_p$.

The top eight companies producing polysilicon, the raw material for fabrication of c-Si PV cells are: OCI Chemical (South Korea), Wacker Chemie (Germany), Hemlock (USA), GCL Solar Energy (China), REC (Norway), MEMC (USA), LDK Silicon (China), Tokuyama (Japan).

The top eight manufacturers account for 64% of global production in 2011, equivalent to 184,922 tons, enough to supply production of c-Si PV cells of around 27 GW/year.

The top 20 companies in 2011 in the c-Si PV cell sector account for 59% of the world production; of them, 11 are from China, including Suntech Power which was the first to surpass the target of 2 GW/year; the top 6 have the production exceeding 1 GW.

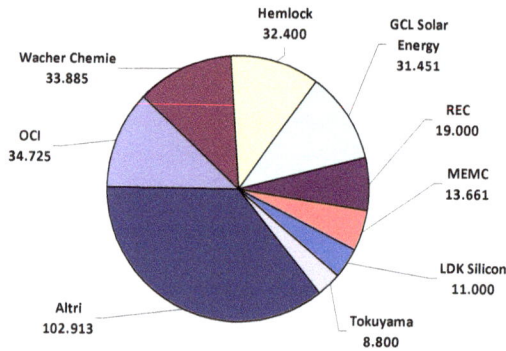

Figure 14.d - Global production of polysilicon in 2011 totalling 287,835 tons by companie

With an exception of the following five - JA Solar, Motech, Gintech, Neo Solar and Q-Cells – all other produce cells for own module production.

Only Taiwanese companies Gintech and Neo Solar produce c-Si cells exclusively.

The list of the top 20 largest companies from the PV module sector is quite similar to the previous one, proving the consolidated trend of progressive vertical integration: 17 companies are present in both lists, and Suntech Power from China leads this sector as well with over 2 GW/ year.

The following differences are notable: the absence of Taiwanese companies specializing in the segment of cells, and First Solar, ranking second, which specializes exclusively for modules with CdTe thin film with a production rate very close to 2 GW/year.

The fact that five companies have production rates higher than 1 GW/year, as compared to only one in 2009, demonstrates that for competitiveness in this sector, the scale is as important as technological innovations.

Among the thin-film module manufacturers, apart from First Solar, the list includes only Sharp with its tandem modules with a silicon thin film. Sharp invested in Italy, along with STMicroelectronics and Enel Green Power, establishing the association 3Sun, that manages a new plant for manufacturing modules using silicon thin film technology.

Located in Catania, by the STM headquarters, the plant started operations from the end of 2011 with a line with 80 MW and it is expected to expand its activities in accordance with trends on European and Mediterranean markets.

In Sicily, there is a company Moncada Solar Equipment producing 40 MW modules of 5.7 m² in size, using Si thin film technology established by Applied Materials –

intended for use by its associated company Moncada Energy to carry out devices for the global market, expected to double its production rate in the next future.

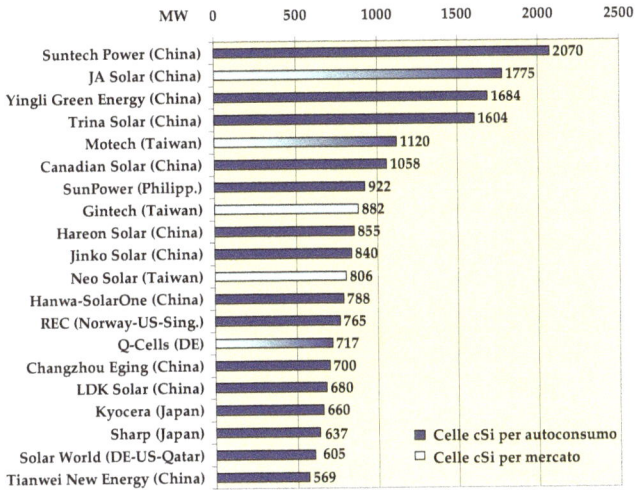

Figure 14.e -Major c-Si PV cell manufacturers in the global market in 2011

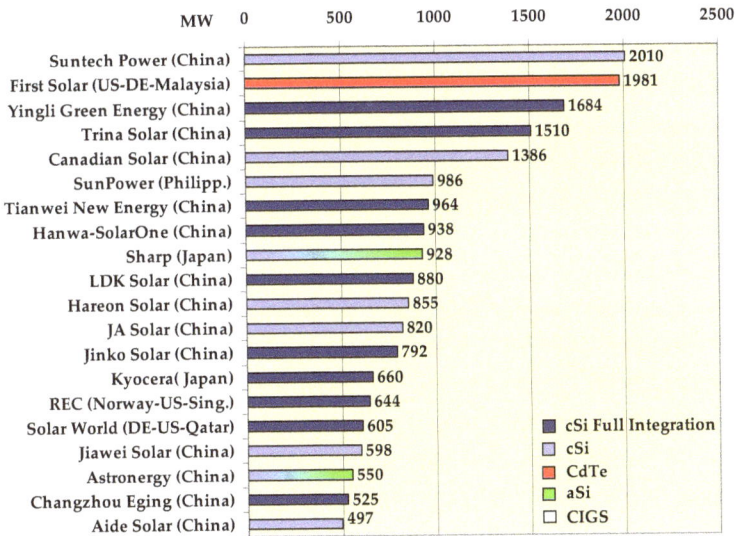

Figure 14.f - Major PV module manufacturers in the global market in 2011

The company Solasonica was largest Italian c-Si module manufacturer in 2011, with around 95 MW and partial integration; its cell production covered no more than 25% of its needs.

Nevertheless, Italy has become one of the major markets globally in the recent years.

In the inverter sector, the number of manufacturers is limited. It is estimated that the top 10 in 2011 accounted for 70-75% of total world production.

By far, the most important company is SMA from Japan, followed by Kaco from Germany and Power One from the USA.

In the province Arezzo, in Italy, Power One has its most important plant that accounted for almost 80% of its total production in 2011.

Company	Country	Production 2011 (GW)
SMA	Japan	7.6
Power One	USA	2.9
Kaco	Germany	3
Fronius	Austria	1.3
Danfoss	Denmark	1

From market pricing evolution, we already pointed out that PV has made remarkable progress in reducing costs, as until recently grid parity still seemed very far away.

It was only a few years ago that PV electricity was four to five times more expensive than fossil fuels.

However, with increases in fossil fuel prices and continuing cost reductions in PV modules, grid parity could occur as early as 2012 to 2013 in sunny regions of USA, Japan and Southern Europe.

Other regions with lower electricity production costs and/or more moderate solar resources may achieve grid parity, as early as 2020.

The capital cost (CAPEX) of a PV system is composed of the PV module cost and the Balance of system (BOS) cost.

The PV module is the interconnected array of PV cells and its cost is determined by raw material costs, notably silicon prices, cell processing/manufacturing and module assembly costs.

The BOS cost includes items, such as the cost of the structural system (e.g. structural installation, racks, site preparation and other attachments), the electrical system costs (e.g. the inverter, transformer, wiring and other electrical connection costs) and the battery or other storage system cost in the case of off-grid applications.

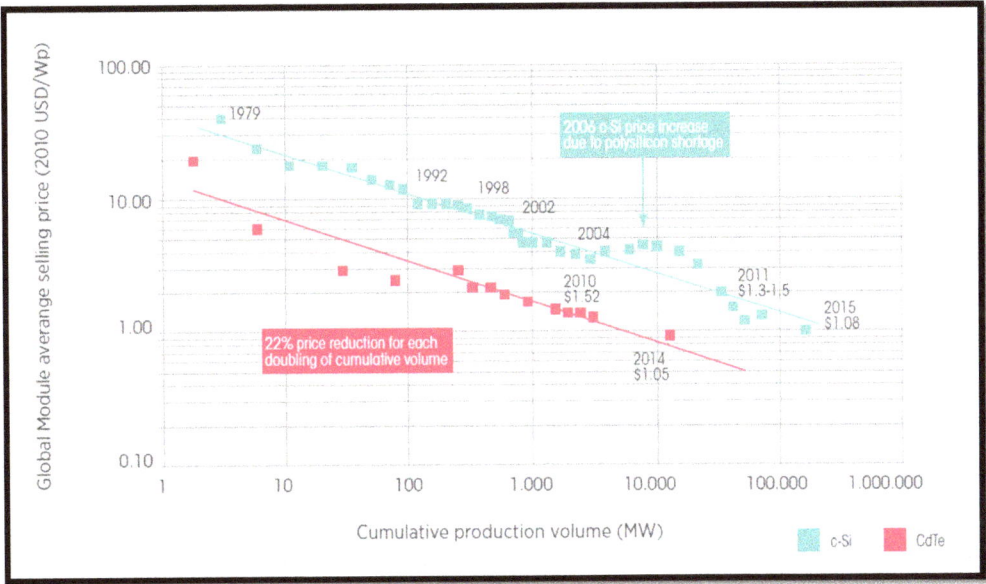

Figure 14.g - The global PV module price learning curve for C-Si wafer-based and CDTe modules, 1979 to 2015

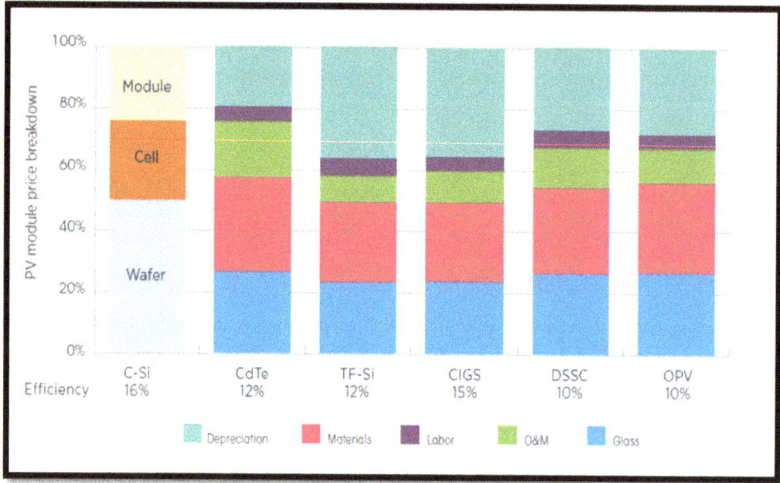

Figure 14.h - Average worldwide PV module price, their cost structure by technology 2010

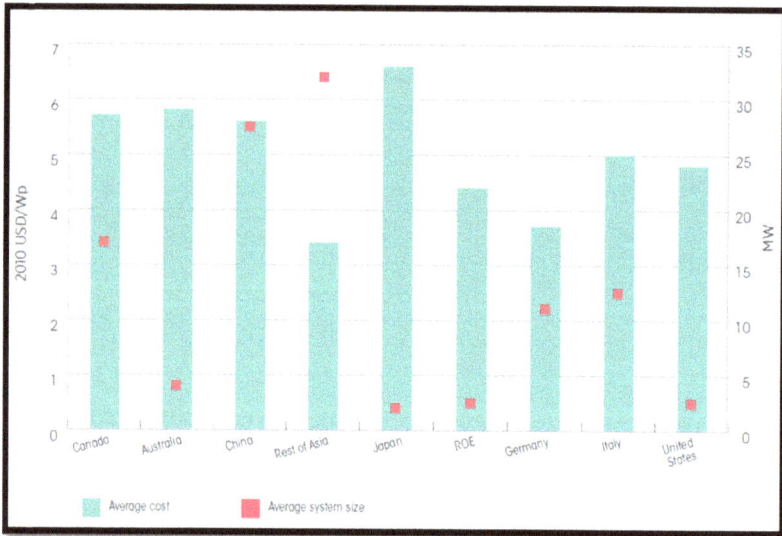

Figure 14.i - Average prices and sizes of large utility-scale PV plants by country 2010

14.2) CSP

Among the other case histories of CSP market approach, we should report the strong government support in Spain: rewarded operators of solar thermal power plants fixed payments at premium prices, sparking a CSP building boom on the southern Iberian Peninsula.

Across the Mediterranean and in other arid or semi-arid regions around the world, plans for solar thermal power plants were moving forward.

Solel, the Israel based company, which manufacturers heat-collecting receiver tubes for parabolic trough-based CSP plants, was a profitable business, as Siemens said at the time, with an order backlog to supply equipment for 15 power plants with a combined capacity of 750 MW.

Studies published in 2009 and 2010 by Greenpeace International, the European Solar Thermal Electric Association (ESTELA) and the International Energy Agency (IEA) went as far as predicting that 84 to 147 GW of CSP could be installed by 2020.

Annual installations would reach up to 6.8 GW in 2015 and 15 GW in 2020.

The IEA, for example in its medium-term renewable energy market report, published in 2012, about 1GW of installations with cumulative capacity reaching 3 GW by end of 2012.

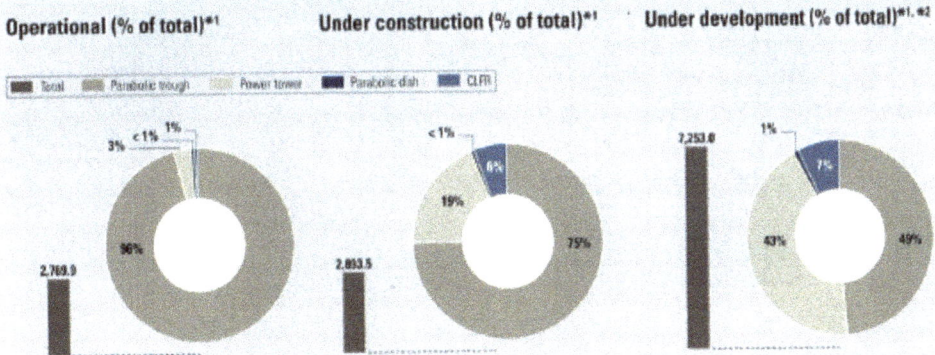

Graph 14.1 - CSP technology shares that are in operation, under construction and under

Through 2017, the IEA now sees average growth of only 1.5 GW per year with total installed generating capacity rising to just 8 GW in 2015 and 11 GW by 2017.

While 96% of all CSP generating capacity in operation today relies on parabolic trough technology, that dominance shrinks to 75% when considering project under construction.

"The technology faces several challenges including increased price competition from solar PV, which has more deployment flexibility; complex environmental permitting; and grid connections", said in the IEA.

For example, wind is expected to lead the way among non-hydro resources, growing to 464 GW by 2017 from 234 GW at the end of 2011.

Instead of becoming the leading supplier of solar thermal systems, as the company anticipated, Siemens has realized less than 300 MW of Solels's pipeline, while its big project under contract with PG&E was canceled.

Three years after, announcing its purchase of Solel and CSP, Siemens unveiled its solar exit strategy, which includes the planned sale of both PV and CSP units.

As an excuse, Siemens claimed that in this environment, specialized companies will be able to maximize their strengths, while in limited markets vertically integrated players often have an advantage this apparently was not the case with Siemens's solar thermal business.

Perhaps this was because of Siemens' focus on parabolic trough technology at a time, when power tower technology is gaining an increasing share of contracts.

In fact, we have to underline that parabolic trough is the most mature technology today and it continues to dominate the market, representing about 95% of facilities in operation at the end of 2011 and 75% of plants under construction by mid-2012.

Another possible reason for Siemens' diminished CSP prospects is that some new markets in North Africa and the Middle East have restrictions on trade with Israel, where Siemens' manufactures its receivers.

Siemens says it will continue to operate its CSP and PV business until they are sold – perhaps even to one of the original owners or one of its competitors. Talks with unnamed, interested parties are ongoing, according to the company, which says it will continue to offer steam turbines for CSP facilities, as well as grid technology and control systems for PV and CSP power plants, after the sale.

The largest of the new power tower projects is the Ivanpah Solar electric Generating Station, a three-phase 377 MW complex under construction in southeastern California.

Currently, CSP investment costs are in the range of $ 4.6 million per MW for parabolic trough plants without storage to nearly $ 10 million for trough plants with storage, according to industry data.

According to market conditions, energy generation costs are in the range of 14 to 35 $cent per kWh – depending on location, plant efficiency, cost of financing and other factors.

Current investment costs for power towers are estimated at $ 6.3 to $ 7.5 million per MW with 6 to 7.5 hours of molten-salt storage and up to $10.5 million per MW with 12 to 15 hours of storage.

Jesse Gary of the US Department of Energy said: "I know I wouldn't be working where I am if I didn't believe CSP has a bright future and I know many of you feel the same way".

Certainly, that is how many of those active in silicon thin-film PV felt about the future of their technology only a few years ago.

The same was probably true for a Siemens' CSP group. For both, that window of opportunity has closed.

The concentrating solar thermal power (CSP) market continued to advance in 2012, with total global capacity up more than 60% to about 2,550 MW.

The market doubled relative to 2011, with Spain accounting for most of the 970 MW brought into operation.

From the end of 2007 through 2012, total global capacity grew at an average annual rate approaching 43% !

Towers/central receivers are becoming more common and accounted for 18% of plants under construction by mid-year, followed by Fresnel (6%) and parabolic dish technologies, which are still under development.

Particularly in developing countries the interest in CSP is on the rise, with investment spreading across Africa, the Middle East, Asia and Latin America.

One of the most active markets in 2012 was South Africa, where construction began on a 50 MW solar power tower and a 100 MW trough plant.

Namibia announced plans for a CSP plant by 2015. Several development banks committed funds for projects planned in the MENA region, where ambitious targets could result in more than 1 GW of new capacity in North Africa in the next few years for domestic use and export.

Saudi Arabia and the UAE plan to install CSP to meet rapidly growing energy demand and reserve more oil for export; Jordan is evaluating possible projects.

In early 2013, Saudi Arabia launched a competitive bidding process that includes significant CSP capacity.

Although activity continued to focus on Spain and the United States, the industry further expanded its focus in Australia, Chile, China, India, the MENA region and South Africa.

There was a general trend of diversification of employment in Spain, the United States and beyond, besides global manufacturing capacity increased slightly during 2012. Falling PV and natural gas prices, the global economic downturn and policy changes in Spain all created uncertainty for CSP manufacturers and developers.

The top companies in 2012 included Abengoa (Spain), a manufacturer and developer, manufacturer Schott Solar (Germany) and developers Acciona, ACS Cobra and Torresol (all Spain), as well as ABB (Switzerland), BrightSource (USA) and ACWA. Saudi-based ACWA emerged as a key player in 2012, with the award of two major projects in alliance with Acciona and TSK (Pakistan), in South Africa and Morocco. Chinese firms have begun to enter the CSP-related component business and are expected to be major suppliers for the foreseeable future.

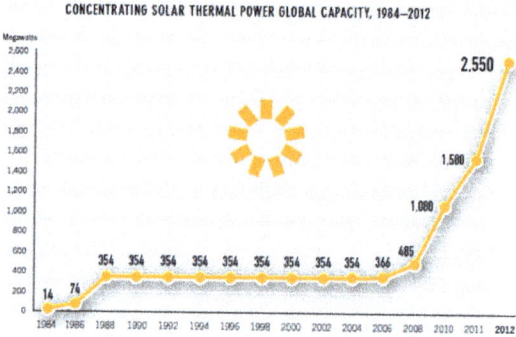

Figure 14.m - CSP Capacity over the years

Thermal energy storage is becoming an increasingly important, feature for new plants as it allows CSP to dispatch electricity to the grid during cloudy periods or at night, provides firm capacity and ancillary services and reduces integration challenges. Molten salt is the most widely used system for storing thermal energy, but other types - including steam, chemical, thermo-clime (use of temperature differentials), and concrete - are also in use or being tested and developed.

CSP prices have declined significantly in recent years, for systems with and without thermal storage.

Although subject to changes in commodity prices, the major components of CSP facilities (including aluminum, concrete, glass and steel) are generally not in tight supply.

To better clarify the main global top players of market, as polysilicon, solar cells and modules manufacturers by capacity, please have a look of the following tables n. 14.n-14.p .

Company	Capacity (Tons)	Country
GCL	65,000	China
OCI	65,000	Korea
Hemlock	43,000	USA
Wacker	33,000	Germany
LDK	25,000	China
REC	19,000	Norway
MEMC	15,000	USA
Tokuyama	9,200	Japan
LCY	8,000	Taiwan
Woongjin	5,000	Korea

Table 14.n - Global top 10 polysilicon manufacturers by capacity

Company	Capacity (MW)	Country
Suntech	2,400	China
JA SOLAR	2,100	China
Trina	1,900	China
Yingli	1,700	China
Motech Solar	1,500	Taiwan
Gintech	1,500	Taiwan
Canadian Solar	1,300	China
Neo Solar Power	1,300	Taiwan
Hanwha Solar One	1,100	Korea
Jinko Solar	1,100	China

Table 14.o - Global top 10 solar cell manufacturers by capacity

Company	Capacity (MW)	Country
Suntech	2,400	China
LDK	2,300	China
Canadian Solar	2,000	China
Trina	1,900	China
Yingli	1,700	China
Hanwha Solarone	1,500	Korea
SolarWorld	1,400	Germany
Jinko	1,100	China
SUNGEN	1,000	China
Sunpower	1,000	USA

Table 14.p - Global top 10 solar module by capacity

15. PV plants and environmental impact

The question of environmental impact, related to installing a PV system and starting its operation, is related primarily to two elements:

- impact to the landscape related to PV system installation and its specific technological configuration;
- issues related to the disposal of system after its useful life is over.

Regarding the first, installing a system of this kind, as any other, is subject to requirements of various bodies for control and spatial planning in as much as it concerns its eventual integration in urban environment.

In particular, for systems larger than 1 MW_p today is required, for example, a special Environmental Impact Assessment (EIA).

In a technical sense, however, more complex is the issue of disposal and recycling of PV systems, as this subject matter has recently undergone substantial changes since the publication of a new European directive on electrical devices disposal.

As a matter of fact, the Recast of the Directive 2002/96/CE on WEEE, of 19 January 2012, included for the first time PV panels into electric and electronic devices subject to special standards of recycling and disposal, provided especially for this category of products, establishing short-term recycling targets of at least 75%, expected to reach 85% in 2018.

All possible scenarios regarding the PV sector development in our country, even the most cautious ones, project that the number of installed PV panels will reach several hundred million units in a matter of years.

Analyses performed by ENEA show that even with these 'conservative' scenarios, as soon as PV systems are decommissioned, the issue will be raised of recycling and disposal of hundreds of thousand tons of glass, aluminum, ultra-purified silicon, precious metals and other materials used for PV panels fabrication on an annual level.

This situation, common to all of the states in the EU, and quite similar to that in other technologically developed counties as well, may lead to serious environmental issues related precisely to the end of the useful life of these products, unless they are addressed in an appropriate way.

The intervention of the European Commission with the above mentioned Directive addresses precisely these issues.

According to the official GSE data (Figure 5.60) in Italy, as of September 2012, the cumulative number of PV systems is 443,747, including both ground- mounted and building-integrated systems.

Each system is made up of a non-negligible number of modules that goes as high as several thousand, in case of PV systems with output power with magnitudes of several hundred kW$_p$.

Peak installed output power totaled around 15.5 GW$_p$ on the same date.

The average lifetime of PV panels used currently is 20 years, with a tendency of a rapid increase to 25 years, already ensured by the most recent technologies. In terms of the quantity of the material in the environment, it can be said that every 100 Wp installed corresponds to around 10 kg of various material: depending on the technology, the number may vary from 3 kg to 12.5 kg.

The above mentioned data is closest to the current technologies.

Hence, in terms of quantity of material to be treated at the end of the useful life of these systems, assuming they are used as long as possible, it can be concluded that if today there are only several hundred tons of material per annum from PV systems installed from 1990 to the beginning of 2000s, than in the period from 2025-2030 there will be hundreds of thousands of tons per year of material to be treated; also taking into account the global growth of the sector in recent years, with installed output power growing several GW$_p$ per year.

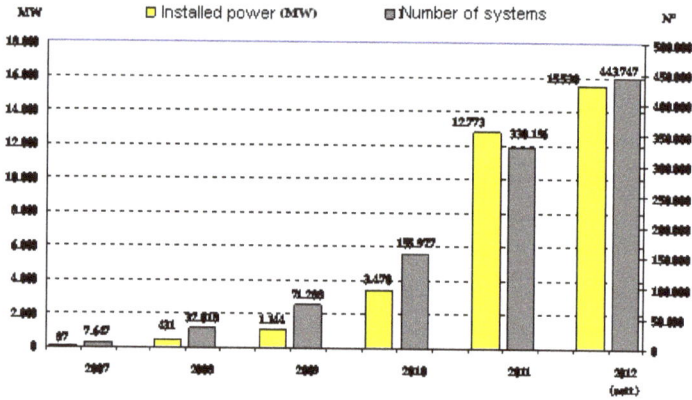

Fig. 15.a – Cumulative output power and total number of PV systems in Italy

From the point of view as is the employed technology, in Italy, as we saw before in the rest of the world, over 95% of installed power is accounted by c-Si or poly crystalline Si, while remaining 5% are panels with amorphous Si thin film (2%) and with CdTe (3%).

It is possible that in the next years this ratio changes slightly, but the percentage of c-Si will remain higher than 90%.

Silicon panels (both crystalline and amorphous) do not contain substances harmful or toxic for the environment, with an exception of minimal amounts of lead contained in the soldering paste that was used in 1990s, and then gradually substituted by pastes more acceptable with respect to the environment.

On the contrary, PV panels based on CdTe contain up to several mg/W_p cadmium. Since the global leader among CdTe panel manufacturers (First Solar) guarantees complete recycling at the end of the life cycle of their products, withdrawing them directly from the market, in the following text it will be dealt only with recycling technology of silicon modules.

Table 5.15 gives basic elements (percentage of total mass) for a silicon based PV panel. Ranges are related to development of specific technologies in course of the past 20 years.

It is therefore easy to observe that, for this type of product, already a simple disassembling of the panel and recycling glass and aluminum would meet the requirements provided by the new EU Directive, at least for a short period.

In practice, the process of recycling and disposal of PV systems is performed in two separate stages:

- Stage 1 – disassembly and transfer of the system to the storage site for recycling (panels and inverters) and to the landfill (the remaining building materials). The owner is responsible for this stage of the system. Costs vary and depend, among others, on the scale of the system. Even though there are not much data available: some scientific Institute, like ENEA for Italy, assesses that it ranges between 400-600€/t.

Table 5.b – Percentage weight of the materials constituting standard c-Si modules

Material	Module c-Si or m-Si (%)
Glass	65-75
Aluminum (frame)	10-15
Plastics (EVA, Tedlar)	5-10
Silicon, Metals (Ag) for cell connections, Pb (soldering)	< 3

➕ Stage 2 – collecting and transport to the recycling site. Parts are sent to various recycling sites where, effectively, disassembling, recycling and recuperation, especially of PV panels, is performed. Costs at this stages of the process are born by manufacturers who usually join into consortiums to provide correct recycling. An example of a voluntary agreement with this purpose, among manufacturers from this sector is PV-Cycle, an association established in Germany as early as in 2007, that has 94 members today, of which 81 are module manufacturers, representing in total 85% of the European PV market. The first cost assessments indicate numbers around 200-250 €/t, including costs of disposing non-recyclable material, such as plastic parts of modules.

From the technical point of view, the only substantial process developed today for recycling silicon PV panels is the 'Deutsche solar' process and its derivations (*European Patent n° 893250*).
The process is based on thermal treatment carried out on high temperatures (around 600 °C), intended for decomposition of organic material used for keeping modules together.
It is followed by disassembling and recuperation (with a 100% yield) of metal parts, such as frames or glass.
Silicon parts, cleaned by chemical methods are reused as a raw material for producing silicon ingots.

16. Polygeneration

The hybrid photovoltaic-thermal technology, or solar cogeneration, is based on using a modular element connecting PV cells and the solar thermal collector forming a unique system for simultaneous conversion of solar radiation into electrical and thermal energy.

The hybrid module may be flat (Photovoltaic/Thermal, PV/T) or concentrated (Concentrated Photovoltaic/ Thermal, CPV/T).

Flat modules employ the conventional c-Si PV cells to collect solar radiation or thin film PV cells, coupled with a panel where liquid or air circulate to collect and convey heat corresponding to a part of absorbed solar energy that has not been converted into electricity (Figure 5.61.)

The total efficiency of commercial products is higher than 50%, with over 700kWh thermal energy on an annual level, generated by a m^2 area exposed to the sunlight (in south Italy there are 1600 kWh/m^2 available annually), with a ratio between generated electricity and heat of 1 to 4 (Figure 5.62).

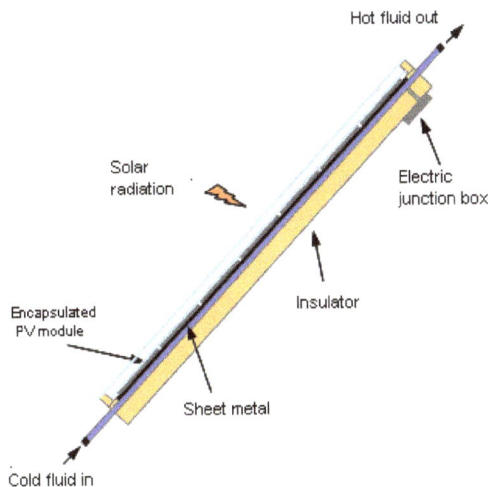

Figure 16.a – Diagram of a hybrid flat module PV/T

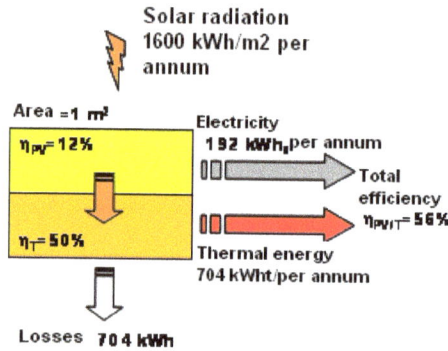

Figure 16.b – Energy flow in a PV/T flat hybrid module

The supplied heat has temperatures up to 60 °C, convenient for residential purposes and households, serving both for sanitary water and heating pools, as well as for winter air-conditioning in combined systems with heat pumps and floor heating.
The advantages of hybrid PV-heat modules of the flat type essentially lay in:

    a. better way of harnessing of incidental sunlight, smaller occupying surface and a total cost lower than the cost of two separate systems, PV and heat, able to provide an equivalent amount of energy;

    b. higher PV generation as a result of lower temperatures of the module in the summer months;

    c. simple architectural integration into buildings.

Hybrid modules may also have a concentrator (CPV-T) and, in that case, direct solar radiation is concentrated through a mirror or a lens onto the collecting surface, covered by PV cells connected in series and supplied with a cooling circuit for collecting and conveying heat, corresponding to the portion of solar energy that is not converted into electricity (Figure 5.63).

The optic concentrator and collecting surface needs to have a moving system to become aligned with the Sun as well as the ratio, between the concentrator area and collecting surface, is equal to the concentration factor, in commercial systems generally below 100X.

The CPV/T systems also provide electricity and thermal energy simultaneously, in the ratio of around 1:4 (Figure 5.64), but they are convenient for locations characterized by high direct radiation (south Italy, countries in the southern part of Mediterranean etc.).

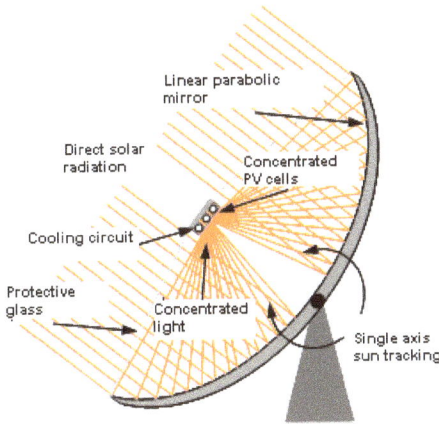

Figure 16.c – Diagram of hybrid PV/T module with a concentrator and energy flow in a typical system.

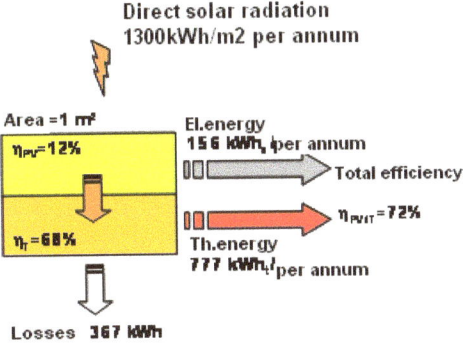

Figure 16.d – Energy flow in a hybrid CPV/T module

Another advantage is the possibility to use the PV cells that have high-efficiency and/or are able to withstand higher temperatures (e.g. multi-junctions with GaAs); it allows achieving higher levels of overall efficiency, and especially, generating heat with higher temperatures (60 < T < 130 °C), sufficient to supply the systems for conditioning with absorption cooling machines and thus to create a 'three-generation system' – electricity + heat + cooling. This can be applied not only in hotels, residential and commercial buildings, shopping malls and sport facilities, but industrial and agricultural sector as well.

The most common configuration of a CPV-T system is Parabolic-Trough with parabolic linear mirrors, a single-axis tracking device, with a low concentration factor (10-30X).

The reason why these systems are so scarce should be sought in their high complexity and higher O&M costs, in the fact that it is not easy at all to integrate them into buildings, and in the lack of specific regulations resulting in a difficult access to incentives.

17. New frontiers of R&D and key opportunities

17.1) Solar chemistry

Solar chemistry deals with activities related to thermo-chemical and photo-chemical processes harnessing solar energy.

Solar energy can, in fact, be converted into solar fuels that can be stored for long periods of time and transported over long distances; it can be used for powering energy-intensive processes for the production of materials at high temperatures; it can also be used for the treatment of polluted air, water and soil or to recycle waste materials.

These processes are still in the research and development stage, but they have a considerable potential in the sense of applications and contributions to solving energy and environmental problems that occur on a global level.

The main processes in the focus of the current research can be classified into three groups:
-   Processes producing energy carriers,
-   Processes producing chemical intermediates (commodity chemicals),
-   Processes de-toxicating and recycling waste materials.

Processes to produce energy carriers. Numerous processes where solar energy is integrated into chemical compounds include the solar reforming of natural gas, biogas and bioethanol to produce hydrogen, the reduction of metal oxides, thermochemical processes for water splitting, solar conversion of carbon compounds, solar cracking of hydrocarbons, solar heat pipe based on e.g. on reforming of methane or ammonia dissociation.

Processes to produce chemical intermediates. – A vast number of chemical intermediates that can be produced utilizing solar processes include, in the first place, hydrogen, synthesis gas, carbon nanotubes, fullerenes, cement, lime, intermediates for the second and third generation biofuels.

Processes to detoxicate and recycle waste materials.

A relevant example of this type of technology include photo-catalytic processes for purification of water heavily polluted by insecticides and fungicides, difficult to treat in normal treatment facilities, gasification process of the sludge from facilities treating urban and industrial wastewater, high-temperature processes of waste material recovery.

Italian excellence of applied R&D needs to be mentioned. Under progress research, over the last few decades, showed many R&D activities in the above mentioned area were carried out in various European countries such as Spain, Germany, Italy, France, Greece, but also in Israel, the USA and Australia.

After a period of stagnation, intensive R&D activities in solar chemistry were launched in Italy in the last decade, especially by ENEA, through long-, medium- and short-term programs.

The efforts were focused on processes with a long term horizon such as thermochemical processes for producing hydrogen from water, in particular the sulfur-iodine process, the process based on manganese ferrites, and the photo-thermochemical process with ammonium sulfur. Later on, the focus shifted on the class of processes for production of hydrogen and solar fuels, using in part fossil fuels, such as natural gas or renewable sources including biomass, biogas or bioethanol and concentrated solar power.

These latter processes, called transition processes, have prospects for industrial applications much closer than the processes using only water, that are currently in the laboratory- or pilot stage.

The goal is to produce hydrogen and biofuels at a cost similar to current industrial processes, but with substantial benefits in terms of saving fossil fuels and reducing climate-altering gases.

Figure 17.a briefly indicates various processes, the subject of the ongoing R&D activities at different international R&D agencies, like the Italian ENEA.

This type of processes usually allow integrating a large portion of solar energy into chemical processes that in normal conditions consume a significant amount of energy for operation, thus allowing for significant primary energy savings and reductions in emissions of climate-altering gases.

One of the processes mentioned above that will be described here in detail is the natural gas/biogas steam reforming, at a more advanced stage of development and demonstration compared to others, close to industrial application.

The first, METISOL project, is financed by the Ministry of Environment, with the participation of ENEA, Fiat Research Center, Energy Company from Lucca, various universities and small and medium enterprises.

The objective of the METISOL project is the production of mixtures of methane and hydrogen for vehicles using solar steam reforming (see in Fig. 17.b a view of the experimental device METISOL).

The second, COMETHY project is a European project financed by Seventh Framework Program, coordinated by ENEA, with a participation of research institutes and European and Italian companies. COMETHY has a goal to produce pure hydrogen utilizing solar steam reforming of various raw materials.

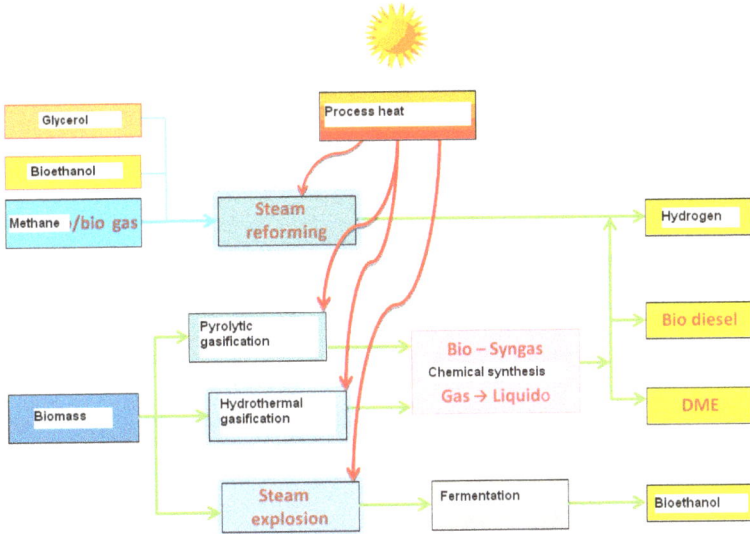

Figure 17.a – Diagram of chemical solar processes at the R&D stage

Figure 17.b – METISOL project: outlook on the solar collector and heat storage/reactor

The diagram of the process of both projects is given before and is applicable to fossil fuels, e.g. or renewable sources - biogas, bioethanol, glycerol.

The heat produced by the solar concentration facility provides the energy required to heat the agents and to maintain the steam reforming reaction.

The saving of primary sources with the proposed solution is between 30-50% and more, depending upon the performance of the traditional process that has been replaced.

Figure 5.68 shows the trend of saving depending on the performance of the process.

The main feature of the solar chemical processes developed is based on the use of the technology of solar energy storage utilizing molten salts, a mixture of sodium and potassium nitrate, which allow efficient and relatively inexpensive storage of energy collected by the solar installations, in large tanks at a temperature of over 550 °C.

This solution allows to overcome the problem of intermittency and unpredictability of solar energy source and to provide continuous and constant supply for the fuel production, with two key benefits: considerable improvement of the system reliability and reducing the amortization time of production facilities that can operate with a very high number of hours per year.

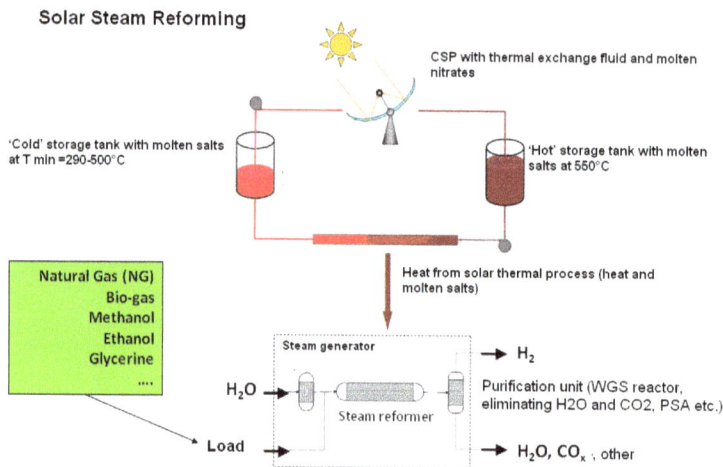

Figure 17.c – Diagram of the solar steam reforming process in a plant

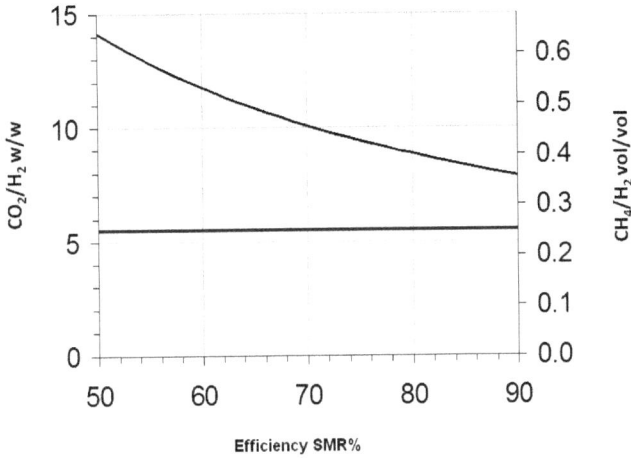

Figure 17.d – Tendency of reducing consumption achievable by solar steam reforming

This technology can be applied by a range of plants, from small to medium ones, decentralized on the territory to large industrial facilities, depending on their needs and characteristics of the area and the industrial environment, with a high degree of overall reliability and a perspective of reducing costs, that would, in the coming years, allow full competitiveness with conventional sources.

17.2) R&D

The electricity generated from solar radiation using the PV effect is, as we saw in the sections above, already a mature and highly reliable technology.

The main components have a quarter century warranty period even in extreme operating conditions, and there is an important research segment principally related to product innovation, i.e. improvement of the commercial product.

As this research often has a very short time horizon, it is carried out primarily in the production facilities of the companies, at least some of which have been mentioned above.

In addition, there are research activities with a longer time horizon, often considering completely different approaches to generating electricity from PVs.

For convenience, these activities are classified according to the PV components: cells, modules, BOS (Balance of system) etc.

The following Section deals with some of these activities following this type of classification.

17.3) New Cells

This topic is frequently discussed by making use of the concept of 'new generation cells', referred to for the first time in the paper by Martin Green published 10 years ago.

In this paper, the development of PV systems is assumed to be based on three various generations of technology: *first generation*, comprising c-Si cell modules, costing around 3 €/W_p and with an average conversion efficiency of 15%; *second generation*, thin film systems (with a-Si, CIGS, CdTe etc.), costing around 0.75€/W_p and with an average conversion efficiency of 10%; and *third generation*, costing approximately as much as second generation technology, but with a much higher efficiency, around 50% on average.

According to the great scientist's vision, the third generation PV systems will finally be relieved from government incentives of any kind, direct or indirect, also eventually become an energetic alternative in a real sense.

Even though Green never discusses this explicitly, it is important to emphasize that the notion of generation involves a hypothesis of the end of silicon PVs and that it will be replaced, first by the second, and then by the third generation in a successful sequence.

Ten years later, however, the scenario has undergone a radical change for the better, compared to what was expected.

Even today, the c-Si PV technology firmly holds its market share of over 87%, based on the costs ranging between 0.6 €/W_p and slightly above 0.8 €/W_p and its conversion efficiency close to 20%.

With such numbers, "outdated" silicon is positioned between 2nd generation and 3rd generation, yet to be achieved in the future.

As for thin film technology, only CdTe has lived up to the expectations regarding 2nd generation cost and efficiency, accordingly awarded by a solid market share of 6% in 2011.

All other types of thin film fall far behind the forecasts, as a result of low efficiency or high costs, or for both reasons.

As far as 3rd generations is concerned, there are various approaches, all of which are still at the research stage.

Already feasible today are multi-junction cells, such as those based on III-V compounds based on germanium, of the type GaInP/GaInAs/Ge, where each layer is optimized for increased absorption of a certain portion of solar specter; effectively, their efficiency has already exceeded 40%[26] (43.5% Solar Junction's cell).

Table 17.e –Record efficiency of cells and mini modules measured in standard conditions (solar specter AM1.5, 1000 W/m², T=25 °C)

Technology	Efficiency (%)	Area (cm²)	Voc (V)	Jsc (mA/cm²)	FF (%)	Testing center and date	Owner
Si (crystalline)	25.0+/-0.5	4.00	0.706	42.7	82.8	Sandia (3/99)	UNSW PERL
Si (multicrystalline)	20.4+/-0.5	1.002	0.664	38.0	80.9	NREL (5/04)	FhG-ISE
GaAs	28.8+/-0.9	0.9927	1.122	29.68	86.5	NREL (5/12)	Alta Devices
GaAs (multicrystalline)	18.4+/-0.5	4.011	0.994	23.2	79.7	NREL (11/95)	RTI
InP	22.1+/-0.7	4.02	0.878	29.5	85.4	NREL (4/90)	Spire
CIGS	19.6+/-0.6	0.996	0.713	34.8	79.2	NREL (4/09)	NREL
CIGS (mini module)	17.4+/-0.5	15.993	0.6815	33.84	75.5	FhG-ISE (10/11)	Solibro
CdTe (cell)	17.3+/-0.5	1.066	0.842	28.99	75.6	NREL (7/11)	First Solar
Si (amorphous)	10.1+/-0.3	1.036	0.886	16.75	67.0	NREL (7/09)	Oerlikon Solar Neuchatel
DSSC	11.0+/-0.3	1.007	0.714	21.93	70.3	AIST (9/11)	Sharp
DSSC (mini module)	9.9 +/-0.4	17.11	0.719	19.4	71.4	AIST (8/10)	Sony
Organic	10.0+/-0.3	1.021	0.899	16.75	66.1	AIST (10/11)	Mitsubishi Chemical
Organic (mini module)	5.2+/-0.2	294.5	0.689	11.73	64.2	AIST (3/12)	Sumitomo
InGaP/GaAs/InGaAs	37.5+/-1.3	1.046	3.015	14.56	85.5	AIST (2/12)	Sharp
a-Si/nc-Si/nc-Si	12.4+/-0.7	1.050	1.936	8.96	71.5	NREL (3/11)	United Solar
a-Si/nc-Si (mini module)	11.7+/-0.4	14.23	5.462	2.99	71.3	AIST (9/04)	Kaneka

Unfortunately, this kind of system has a very high price and still is reserved only for the CPV market, which, generally, cost today well above $5 \ €/W_p$.

A real step toward 3rd generation can probably be made utilizing systems able to use the portion of solar specter, that typically becomes thermally dissipated.

An effect, that can be used for this, is the Multiple Exciton Generation effect, where a part of the energy that is not used for recombination, not caused by radiation and induced by a high-energy photon, is not lost through heat, but reused as another source of electricity (highly complex mechanism based on a kind of a resonant Auger process in nanostructure materials.).

It should be noted though, that while the reliability of silicon PV systems, and partially, thin-film systems, is guaranteed for a period of 25 years, various approaches to third generation are still at the laboratory level: we do not know anything about their practical applications on a time scale comparable to the above mentioned one.

It is clear from the above that the notion of a 'new generation cell' should be used with caution and that it has the same value in industrial and market context because it considers the cost per W_p as a main unit of measure.

However, there are also more basic research approaches in which the cost is and should be considered only a marginal element.

This is, for example, the case of organic cells. All types of PV systems utilizing, in some way, the carbon compounds as the active material are included in this class. Thus, organic cells are cells based on aromatic compounds such as phthalocyanines, polymeric compounds, e.g. of the family of polythiophenes, type P3HT or dye of the pyridine-ruthenium type, used in Graetzel, cells commonly known as DSSU cells.

It should be said that even though the class of membership can be considered as unique, the structure of the active device and its operating mechanism vary from case to case.

Moreover, the large diversity of available materials and process technologies that can be used allows multiple designs of the system, even within a single product family.

This class of devices is also distinguished by its operating mechanism, clearly different from Si cells, cells with thin film or CPV discussed above. In fact, in the case of a cell in which the active material is a conjugated polymer such as P3HT, from the viewpoint of electronics, the limited extension of chains of the conductive polymer (hundreds of nm) creates a material with a band structure similar to semiconductors, but with a gap depending on the particular chemical composition of the polymer itself. From this definition three things can immediately be deduced:

1) order of chains influences the properties of the electrons;
2) valence bands have a variable Homo-Lumo distance (band edges);
3) due to the type of the chains (in addition to its chemical structure), a bulk of these compounds is very reactive with water and oxygen compounds.

From the point of view of interaction with light, the optical absorption is very high. Generally, the problem is to extend to the red-NIR region, and this is one of the main directions of the research. Interaction with light creates an exciton (in actual reality, it creates a myriad of particles – excitons, bi-excitons, solitons, etc., but a pair of excitons is the most common and most widely utilized in the PV systems).

The excited pair lives short, just for several nanoseconds: if in this period an external force manages to separate them, bringing an electron into the conducting band, the effect may be used to generate photocurrent, and with the appropriate design of the device, also to generate electromotive force.

Another type of architecture frequently reported in the literature is that of simple hetero-junction that still allows for timely separation of charges.

At this point, the pair is separated, and hole and electron both generate photocurrent in the device.

The mechanism of charge transport, however, is not diffusive, typical of inorganic device.

In this case, the mobility is very low (e.g. 10^{-7} cm^2/(Vs) for holes and slightly higher for electrons) and thus the current is dominated by the spatial charge recombination phenomena (SCRC).

Therefore, the electric contact is less important than in the case of inorganic compounds (there is no need to have an efficient contact if the charges reach it too slowly), but the conversion efficiency may be very low as the photo-generated charges 'die' on the way, before reaching the contacts.

This framework allows room to improve the thickness of the active material: the high absorption coefficients allow the use of very low thickness values, of just a few hundreds of nm.

Therefore, to determine the particular transport mechanism, it is important to consider the 'external' electric field that also modifies the mobility, establishing different transport mechanisms for different polarization regions of the device.

The major issue of this type of devices still is their stability over time: in the best of cases, their operating life does not exceed 1000 h (certified).

The Graetzel cells may be considered a special case of organic cells.

In these devices, the function of the polymer is performed by a dye, where an electrochemical reaction, activated by sunlight, generates electron-hole pairs.

The electron is collected by a fine dispersion of nano-particles of titanium oxide and transported, by means of a transparent conductor, to a platinum electrode that allows to close the circuit, collecting another photo-generated carrier.

Even in this type of device, the high porosity of the electrodes is a property of essential importance for the efficiency of the device.

In the area of poly-crystalline thin film cells, based on CIGS (Copper Indium Gallium Selenide), research is carried out in various laboratories, with a goal to overcome the problem of poor availability of indium by developing materials similar to CIS, where indium is replaced by a pair of elements from the II-IV group.

Regarding thin-film devices, it is known that those based on CIGS alloys have shown the highest potential to reach high levels of conversion efficiency, currently close to 20%.

Therefore, on this basis, it reasonable to explore variants of the process and composition in order to avoid the problem of indium.

In early 2010s, IBM introduced a method of depositing an alloy of $Cu_2ZnSn(S,Se)_4$ that gave excellent results.

Alternative techniques for the same material were also proposed in Japan. The material, called CZTS, has a crystalline structure called kesterite.

The process involves the deposition (by sputtering or evaporation) of a precursor film that contains metal elements in exact proportions, which is then heated to around 500 °C in a gaseous environment with sulfur to crystallize CZTS.

The advances by the IBM laboratories allowed a record efficiency of 11.1%.

Cells based on graphene have recently also been developed as one of the new PV devices.

The excellent electrical conductivity and high optical transparency of graphene make it an ideal candidate for producing transparent electrodes and conductors, with important consequences in the field of PVs, in particular to replace metal in the devices based on Schottky junctions.

The graphite has already been usefully employed in this application in combination with a wide range of semiconductors, including Si, GaAs and 4H-SiC.

The Schottky barrier formed at the graphite/semiconductor interface is extremely strong and has numerous advantages compared to the one formed by the traditional metals.

Because of the strong bond between carbon atoms and the relatively small size of that atom, the phenomena of carbon migration in the semiconductor do not occur, thus all rectifying properties of the barrier are preserved.

Besides, unlike metal whose Fermi level is fixed, graphite can be doped and this allows the modulation of the energy levels, and consequently also the levels of Schottky barrier.

Finally, not being a heavy metal, graphite is not toxic.

The replacement of graphite with graphene would provide an additional advantage of overcoming the problem that has so far limited the use of Schottky devices in the field of PVs, i.e. absorption of incident light by the front metal layer: in this case, the high transparency of graphene allows the incident light to penetrate it almost unaltered, thus opening the way for the use of Schottky devices in the field of PV systems.

17.4) Smart Photovoltaics

The term smart PV refers to the implementation of additional electronic, sensing and ICT features to a photovoltaic device.

As we have seen in the sections above, the cost of energy produced by a photovoltaic system depends to a considerable degree on the optimization of productivity at any given moment, and the lifetime of the system.

A smart PV system is designed to prevent the main problems arising from both of these issues. We shall take a look at a few:

> *Fault or shading of a single module* – Typically, the modules in a system are connected in series, within a single string, to the inverter.
>
> If one of the modules fails, the use of by-pass diodes can still ensure the flow of current, albeit at a lower voltage. An alternative solution would be to integrate a level of single module DC/DC converters or single DC/AC inverters.

> *Fault or shading of a single cell* - This problem scales down the problem described above to the level of a single module: the failure of a single cell can be reflected as a failure of the corresponding module.
>
> The use of appropriate electronics can, in this case, be of help to allow isolation of the defective circuit element, allowing the module to continue operating.

It is evident that both solutions discussed above also apply to the cases of *non-uniformity* among modules or among cells, allowing, for example, the optimization of electricity generation in a PV system and in conditions where the degradation effects in the course of the lifetime of the system did not affect the entire system uniformly.

The integration of appropriate wireless sensors (temperature, current, voltage) at the level of module with appropriate methods of sensor fusion at the level of control station of the system, can easily allow *stability control* for the operation time of a single module, allowing the programming of maintenance activities and consequently also the maximization of productivity.

Finally, the integration of electronic labels will allow a quicker identification of the device and its parts, both for purposes of *control* by administrative authorities and protection *against theft*, as well as for *disposal and recycling* of the system at the end of its life.

A smart system is thus a system that *facilitates the installation*, allowing, for example, the optimization of the choice of inclination and simplifying connection to the grid. In the longer term, ICT functions will ultimately enable the system to operate as a *smart element of the grid*, allowing the optimization of power generation with respect to expected consumption and weather conditions.

The techniques for implementing the above described functions into a PV system are at the prototype stage.

Some products are already available, but we will be able to speak of a proper smart photovoltaic system when the elements of the electricity generation, sensing elements related to the monitoring of the performance and functionality, and finally, electronic elements related to the management of various operational situations, become elements derived from a single manufacturing process.

17.5) Photovoltaic systems and smart grid

The strong presence of grid-connected power generation facilities from renewable sources (e.g. wind, solar) has brought a number of issues to the attention of operators in the electricity sector.

Closer attention to such issues will require a real revolution in operational modalities and management of the grid.

The distribution grid, with the presence of distributed generation systems, has radically changed its status as a passive grid that, in a unidirectional flow of energy and operation, performs its task between the centralized power generation and the end user.

Today, the possibility that the grid assumes an active role because of the distributed generation, instability and limited programmability of these sources, has resulted in focusing the efforts of research and innovation on a new concept of the grid.

This new concept, based on viewing the grid as a set of patterns and active bidirectional devices simultaneously installed on the grid, in power plants, and ultimately on consumers, in order to fully integrate renewable sources and other functions for power quality and safety of the grid, is the concept of *Smart Grid*.

The integration of solar technologies in the grid, in the first place, has played an important role of providing support at peak demand, particularly in the areas with warm weather and increased power demand (e.g. as a result of greater air conditioning in summer), when the peak demand coincides with periods of high solar incidence. The distributor is then able to increase supplying the grid with the clean energy, that, in those hours of supply, may also be more affordable than electricity from conventional sources.

Currently the level of generation by photovoltaics and, more generally, from renewable sources, goes well beyond this supporting role to the network.

In Italy, the great increase in the electricity generation from renewable non-programmable sources or "FERNP", especially wind and solar power as well as the corresponding exponential growth of the companies producing this type of energy, scattered throughout Italy (almost 450,000 plants for PV only), particularly in the southern regions, has been profoundly changing the energy market and is imposing a radical transformation to the distribution grid both in terms of strengthening, and, especially, in terms of additional services.

According to data from TERNA, the installed power of FERNP (over 19 GW) is significantly closer to the grid load that varies over the course of average days between 20 and 50 GW.

In addition, the electricity generation from FERNP during the daylight hours covered in recent months over 16% of demand (as compared to 8% in early 2011) with a reduction in the market price, undermining the production system of some fossil fuel facilities many of which were forced to shut down.

At the same time it raised the price of electricity at night and in the evening when, inevitably, plants powered by fossil fuels come into play with producers eager to make up for economic losses during the daylight hours.

Therefore, it is clear that at present photovoltaics hold a dominant share in the market as a result of the storage system of the excess electricity, generated during the daylight hours, as well as the systems for management of stored energy to enhance postponed consumption.

In short, the variability of these sources, the need for programmability of the supply, safety concerns, couple with the importance of trading on the electricity market, all require the adoption of new architecture, devices and management approaches in order to ensure the supply of electricity characterized by high quality standards, stability and active involvement of suppliers at all levels.

The concepts of smart PV and smart inverter in the PV sector fit exactly into this strategic direction, for implementation of Smart Grids with the general goals to:

- Maximize the integration of electricity generation from renewable sources by facilitating connecting to the grid.
- Stimulate the demand.
- Allow higher participation of end users in the electricity market.
- Ensure the operability of the system even in conditions of grid disruption.
- Establish an appropriate balance of generated and absorbed power at all levels of the distribution system.
- Integrate the systems that are currently separated physically and functionally (energy, transport, telecommunications, environment, land, etc.), aiming at the full implementation of a range of services intended for citizens in accordance with the *smart city* concept.

In order to enable these functions, the systems are supported by ICT platforms, that play a key role relevant for the entire smart grid. ICT systems will be responsible for the measurements of the observable parameters of the grid, transmission and analysis of certain parameters of the system and the management and control at different stages of the operation.

In the future smart networks, ICT systems will be set as a means to strengthen the coordination of all components of the grid in order to increase its robustness and security.

The new ICT technologies will also introduce new rules for energy trading, based on dynamic and flexible tariffs determined in a real time environment.

The combined potential of these technologies open the possibility of introducing portals for the control of energy systems, at all scales, to manage the consumption and generation, automatic purchasing systems, storage and the sale of energy, supply and demand balancing systems, and many others..

In short, a smart grid will be a great interactive infrastructure, a kind of 'Energy Internet', capable of exchanging information in real time and developing an integrated policy of control and automation of the electrical system as a whole.

In Italy, the big players of transmission and distribution grid, supported by universities and research institutes, are performing important initiatives, within the Smart Grid framework.

The first experiences date back to 2006, owing to ENEL, who coordinates with ENEA on the international level, in the area of the smart grid of the EERA (European Energy Research Alliance).

The Interregional Operational Program on Renewable Energy and Energy Saving for 2007-2013 ('POI Energia') supports measures to increase the efficiency, energy saving and generation of renewable sources, including the strengthening and upgrading of the transmission grids of electricity (Priority 2 of the 'POI Energia').

The following programs have been introduced: 'Strengthening Medium Voltage Grids', intended for renewable energy producers, with investments of over 123 million euro for the period of 2010-2014 and 'Intelligent Grids of Medium Voltage' with investments of 77 million euro for the period of 2009-2013.

17.6) Geocoding

Advanced technology and know-how of micro-siting analysis are coming from advanced expertise and computational procedures to abstract detailed info through satellites monitoring and SAR data acquisition.

Currently, the author already developed and applied it for significant international patent to protect such as geocoding methodology for satellite data processing, which may mainly support and accelerate the project development during all kind of energy project: above all, during exporation and micro-siting 3-D model construction phase of renewable source under green-field designing.

To focus on geocoding applications, I should begin from the program language "Interactive Data Language", through that one may select and authomatically take out very important informations from "Synthetic Aperture Radar" or SAR processing.

SAR acquisition and its geocoding need to know several data coming from the final user, as follows: satellite coordinates, perimeter of target and SAR image parameters to be elaborated.

Geographical coordinates can be obtained like latitude and longitude, or north and east of each cartographic projection.

Thus, the main steps of geocoding technics are the following:
- ✓ Calculation of search table
- ✓ Re-sampling of data set through that search table

as the following scheme of SAR geocoding process can show us:

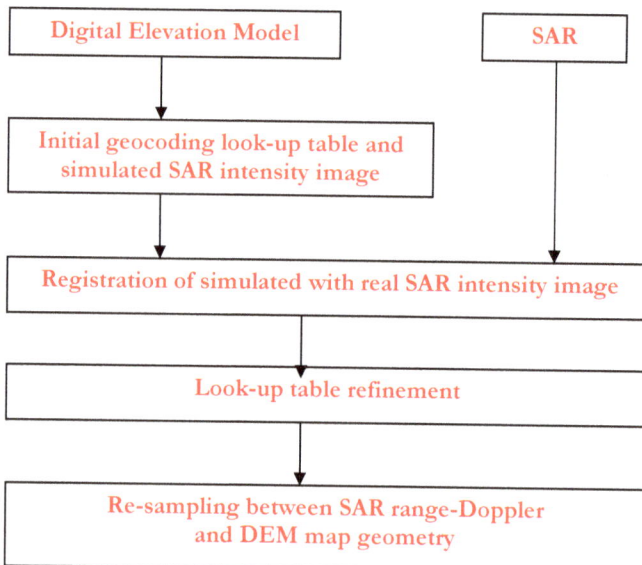

```
┌─────────────────────────────┐        ┌──────────────┐
│  Digital Elevation Model    │        │     SAR      │
└─────────────────────────────┘        └──────────────┘
              │                                 │
              ▼                                 │
┌─────────────────────────────┐                │
│  Initial geocoding look-up   │                │
│  table and simulated SAR     │                │
│  intensity image             │                │
└─────────────────────────────┘                │
              │                                 │
              ▼                                 ▼
┌───────────────────────────────────────────────────┐
│  Registration of simulated with real SAR intensity │
│  image                                             │
└───────────────────────────────────────────────────┘
              │
              ▼
┌───────────────────────────────────────────────────┐
│  Look-up table refinement                          │
└───────────────────────────────────────────────────┘
              │
              ▼
┌───────────────────────────────────────────────────┐
│  Re-sampling between SAR range-Doppler             │
│  and DEM map geometry                              │
└───────────────────────────────────────────────────┘
```

To better clarify how SAR geocoding process is working, the scheme of flow diagram can show it above.

Such as search table holds a corresponding coordinate into geometry range-doppler SAR, for every pixel of geocoded image from cartographic and selected domain.

SAR interferometry or "InSAR" is a key method to approach geocoding process and shows to be a powerfull tool for measuring of surface topography, interesting a renewable energy plant development, as solar farm could be into smart city integration.

Moreover, interferometry study is able to report changes, according from short to long time period, plus additional surface changes.

The relevant performances of interferometric system depend on radar sensor parameters as well as the flight coordinates identification.

SAR acquisition process provides a geometric projection of a radar riflectivity:

- $a(x, y, z)$

from 3-D space into a 2-D system of cilindric coordinate (x, R), as the following equation show us:

$$y(x, R) = \left(\int a(r)R \, d\theta \, e^{-j\frac{4\pi}{\lambda}R} \right) * g(x, R)$$

Where:

✓ $y(x, R)$ means SAR image,

✓ $r = (x, y, z)$ is the position of diffusion single point,

✓ x means the azimuth direction, y is ground-range, z is hight, orthogonal to plan (x, y),

✓ R is the slant-range or distance btw SAR sensor and diffusion point,

✓ θ is the elevation angle,

✓ $r = (x, R, \theta)$ means position of diffusion point in cilindric coordinates.

In the light of new generation of satellites as "TerraSAR-X", "CosmoSkyMed", "Quickbird", "Ikonos"…can really provide huge amount of SAR and optical data with metric or sub-metric high resolution (HR).

That strong device should be specially applied to study urban as well as country and set-aside lands micro-siting, to finally support urban plan, smart cities and grids evolution.

18. Financing

Any industrial project, as a solar energy facility, must be designed, developed (permitted, engineered and constructed) and approved for the final investment step, covering its industrial *Value Chain* and according the following requirements for project financing:

- **Fitting of business environment and conditions**
  1. Bankability from capitals market.
  2. Legal Framework as Energy Law, plus the 1th and 2th level to support RES production.
  3. Regulatory Framework & Market Pricing, as Independent Body and other Authorities that should monitor and regulate the market dynamics and, where it's possible, delivery special incentive mechanisms (ex: Green Certificate, Feed-in-Tarif, Feed-in-Premium).
  4. Power Purchase Agreement or other long term contract that, by law, should necessary go for setting up both technical and commercial terms&conditions to sale power and heat generation to the market operator.
  5. Priority principle of dispatching 100% as green generation.
  6. Preferential role of green-producer (ex: electrical grid connection).
  7. Positive public opinion on green energy.
  8. Local presence with strong network relationship (ex: Institutional & Industrial Stakeholders).
  9. Business / Technology diversification according its business plan.
  10. Industrial sinergy with Gas or Black Power and other energy businesses.
  11. Regional Energy Balance and future scenario, also according the electric generation and interconnection development plans.
  12. Country focus and CO_2 balance to size the CO_2-Debt reduction plan.

- **Key value chain steps**
  1. Project scouting and analysis, meaning:
     - ✓ Resource evaluation,
     - ✓ Grid study,
     - ✓ Land rights.

2. Permitting and design optimization, meaning:
   - ✓ Legal issues,
   - ✓ Contracting,
   - ✓ Pre-Valuation.

3. Financing scheme and contractors of project, meaning:
   - ✓ Project Due Diligence and Valuation,
   - ✓ Financial Model and Business Plan,
   - ✓ DD Reporting,
   - ✓ EPC contractor selection,
   - ✓ Bridge / Project Financing application.

4. Engineering, Procurement, Construction contracting and its bankability.

5. Operation and maintenance contracting and its bankability.

Those special topics are all together assuming the main and final project's target: the solar plant bankability !

Several and flexible options to the above terms depends on both: best practice of project development and the operating business model of the industrial player, as "Make" or "Buy" approaches.

In the case of "make", the deal shall focus on the organic business development method, with a specific team operating in house (or in outsourcing sometimes) to carry out green field project and reach, step by step, all the most relevant milestones, like technical feasibility and design up to permitting process.

In the second one, "buy" approach, industrial and big player likes to scout and short list from the market (local developers) the most advanced and authorized project, in order to reach brown-field projects, under M&A practice, in some specific conditions: "Ready to build" or "Fully permitted" or "Turn-key" plant acquisition after construction phase.

Usually, that one results the preferred way to go to the market by Energy Utilities or big Industry and Investment Funds.

18.1) Risk Profiles

To better identify the most significant risk areas of a solar project development, investment and execution cycles, it is worth proceeding across its value chain steps, as follows:

Category	Task	
Development	Identify site	The most project risks are connected to the development operational phases.
Development	Assess site quality	
Development	External advisors	
Development	Secure ownership rights	
Development	Assess technology options	
Development	Preliminary design	
Development	Permit application	
Pre-construction	Supplier selection	Choosing of EPC and technology suppliers are key points for project financing endorsement and acceptance
Pre-construction	Construction contracts	
Pre-construction	External advisors	
Pre-construction	Project information management	
Project financing	Bank selection	Optimal bankability conditions
Project financing	Technical advisors	
Project financing	Project financing	
Construction	Civil works	Low complexity meaning low risk level of project works, building and commissioning
Construction	Mechanical assembly	
Construction	Electrical works	
Construction	Commissioning	
Operation	Connection to grid	In some cases, that risk could depends on special deadline by Law, but after grid connection all risks are basically insurable.

The best practice of industrial and financial investors shall focus on each specific risk assessment and classification, during all the steps of supply-chain, from solar plant engineering, supply and construction up to commissioning, operating and maintenance phases.

In that way, industry or banks financing the solar project will be able to better clarify all the specific mitigation tools, like insurance program and its amount and characterization, as well as the identification of critical liabilities, indemnity period and deductible targets.

Basically, a preliminary risk matrix and classification should start from the following risk ranking and insurances perimeter:

Insurances based on risk	Insured amount	Indemnity Period	Deductibles
Construction and Erection All Risks (CAR, EAR)	Estimated on contract value basis		
Construction Third Party Liabilities	Capped value depending on Market regulation and Company policy		
Delay in Start Up	Amount depending on the certified power production (estimated by independent opinion)	12 Months as usual	
Marine Transit	To be advised as: max transit value, marine transit split, no. of shoments,…		
Marine Transit Delay in Start Up	Amount depending on the certified power production (estimated by independent opinion)	12 Months as usual	
Operational All Risks	Full Replacement Value		
Operational Third Party	Capped value depending on Market regulation and Company policy		
Business Interruption	Capped value for 12 Months, depending on certified power production (estimated by independent opinion)	12 Months as usual	

To focus on the insurances for Construction Phases, firstly one needs to face a bankable selection process of experienced EPC as the Contractor All Risk.

The Contractor shall stipulate a Contractors All Risk insurance policy (CAR) with an Insurance company of international relevance for an amount at least equal to the total value of all works under contract.

CAR covers indemnity in respect of the cost of reinstatement, replacement or repair of physical loss destruction or damage to permanent and temporary works, undertaken in relation to the construction of the development, including unfixed material and goods, whilst in storage and in transfer overland.

This covers construction, testing, commissioning and handover phases.

In front of that scheme, the "Estimated Contract Value" shall be the same "Full Replacement Value" or the CAPEX budget as well.

Besides, the Contractor shall stipulate a Delay in Start-Up insurance (DSU), which indemnify against loss of revenues suffered and/or additional financing costs incurred during a defined indemnity period, in consequence of delays in the completion of construction of the works, resulting from an event under the CAR insurance.

The Indemnity Period will be usually equal to the total Construction Period, for a total insured amount and a maximum deductible will be around 21 days.

The Contractor shall stipulate an Insurance to provide cover of all risks of loss or damage for plant equipment and materials, that travel by road and over land (in case not already included in the CAR) and by air or sea.

The Contractor shall stipulate a Third Party Liabilities insurance also, to indemnify against all sums (including Claimant's costs and expenses), which the insured shall become legally liable to pay as damages attributable to the Contractor and/or other persons acting behalf of the Contractor (whether under contract or otherwise), in respect of or consequent upon:

- Death of or bodily injury to or illness or disease contracted by any person (other than the employees of the insured company)
- Loss of or damage to property (other than the property insured under the CAR coverage), as
  - Damages to environment and property of third parties on which and/or in which the works are performed;
  - Damage to existing structures;
  - Damage to property of third parties placed in areas where the works are being performed;
  - Damage to adjoining properties;
  - Damage to underground pipelines or other facility;
  - Damage and / or theft of furniture and equipment of the Employer.

With specific reference to the solar plant operation, the Contractor of a global O&M service agreement shall stipulate an insurance policy covering physical damage triggered losses, during the operating period, starting from the date of the minute of final acceptance test.

The insured amount will be the Full Replacement Value of constructions and equipment provided by the Contractor.

The Contractor shall stipulate a Third Party Liabilities insurance during the operating period as well, starting from the date of the minute of final acceptance test.

As additional requirement, the Contractor, which shall provide a copy of such policies to the Employer's Engineer for approval at the time of takeover of the areas, shall comply with all requirements of the policy also in regard to any claim for damages and procedures associated with the settlement. In case of default, the Engineer may withhold any payment due to the Company until the submission and Engineer's approval. Amounts withheld on that basis shall not bear interest for the Company.

An other industrial classification matrix could be arranged and diversified into the following risk areas: financial, physical, third-party, deductible.

Financial Risk	Delay in Start-Up (DSU)	Marine Transit/Cargo DSU	Business Interuption (BI)
Physical Risk	Marine/Inland Transit/Cargo	Erection All Risk (EAR)	Operational All Risk
Third Party Risk	Primary Third Party Liability (TPL) Policy. Excess TPL for higher limits if required		
Policy Deductible	Retained Risk		

18.2) Best practice of FIDIC standards

To introduce FIDIC's system of contracts, we should remember that FIDIC's Contracts Committee produces specific forms of contracts for civil and power engineering projects in order to globally standardize terms and conditions basis.

Such a final purpose of FIDIC's contracts is focused on the best definition of relationship through the agreement between the parties as well as to identify and fairly share the risks during engineering, procurement, construction and management, or EPCM, phases between the contractor and the project employer.

Thus, international FIDIC standard applications aim to clarify and allocate each risk to the party that is best able to bear and control it across the steps of engineering, procurement, construction and management.

With respect to solar power project and plant realization, we need to go deeper into different FIDIC forms, depending on specific type of contemplated projects:

Due to the most relevant renewable source market approach, under Turn-Key and LumpSum EPC contractor plus project financing of solar deal, SILVER BOOK is the proper FIDIC standard, that should be always adopted.

In that case, Silver Book conditions push the contractor to provide a completed generation facility to the employer, meaning the plant is ready to be operated at "the turn of a key" and, accordingly, start-up of cash-flows, from project financing side.

In front of the strongest bankability terms, as project-financing no-recourse, the Silver Book is used as an heavy guarantee, where the certainty of price and completion date are important, as well as that's a significant market indicator of regulatory and legal frameworks stability.

I need to emphasize that is of critical importance in such projects not only for the projects to be delivered within time and cost constraints, but also to be delivered so that it is capable of meeting its designed production and output levels.

Performance of the asset is particularly key in those turnkey projects, funded through project financing.

Lenders' security is dependent largely on the ability of the completed facility to operate and generate revenue.

The extent to which risk is allocated to the contractor under turnkey arrangements will depend upon a range of other factors, including the availability and strengths of quarantees from the project's sponsors.

Where a sponsor will not provide any, or only a limited form of, completion guarantee to lenders, this obviously increases the need to allocate completion risk away from the sponsor.

In these circumstances, the obvious candidate for the risk, given that it will be in the best position to manage it, will be the turnkey contractor.

Therefore, turnkey contract is the means by which the risk is allocated.

The key risk in any solar or different power construction project is completion risk, meaning that the works may not be completed:

1. Within the agreed lump sum price; or
2. Within the agreed time scale program; or
3. To the required performance quality.

In a turnkey agreement, it is the contractor who has responsibility for and control over each of these elements of completion risk.

Large projects will frequently involve a number of turnkey contractors undertaking different parts of the overall project, each according to its own specialist skills.

Such approaches are also common in process engineering projects, where the output may be energy generation from both renewable and conventional, besides water treatment, petrochemicals or natural resource processing.

From the other side, the Silver Book application may also be a relevant project tool for privately financed on "Build-Operate-Transfer" basis (BOT) assets.

In that business case (ex: emerging market Countries), the employer shall take all the responsibility for design, construction, maintenance and operation of a project, with changing of risk sharing method with the contractor.

18.3) Power Purchase Agreements (PPAs) in the Solar Industry

by

Avv. Lorenzo Parola

(lorenzoparola@paulhastings.com)

<u>Definition, Standard Risk Allocation and Rationale</u>

A power purchase agreement (PPA) is a typical long-term "take-or-pay" sales agreement between a Generator (usually an independent power producer – IPP) and an Offtaker (usually an utility) whereby the latter agrees to purchase over a certain period of time a minimum agreed quantity of the IPP's plant power output at an agreed price.

If the Offtaker fails to purchase the agreed quantity, it is nonetheless obliged to pay the agreed price. This enables the Generator (which is usually a special purpose vehicle (SPV) incorporated in the context of a non-recourse project financing) to get a guaranteed cash flow aimed at obtaining and repaying the debt raised for the construction of the power plant, covering operating costs and remunerating the equity invested by the SPV's shareholders (ie the Sponsors of the solar power plant project).

A PPA related to a portfolio of plants is sometimes referred to as a "virtual power plant" agreement (VPP).

In a bankable PPA the Generator must benefit from an assured cash flow subject only to availability and performance of its power plant and the Offtaker complying with the terms of the PPA. Therefore, the Generator carries no (or very limited) power price and volume risks. This is common to all PPAs either related to conventional (ie fossil fuel-fired) plants or renewables plants (including solar plants). However, certain important differences apply to PPAs for solar plants in that:

- given the absence of a fuel supply agreement a solar Generator does not suffer from either fuel supplier's default risk or fuel price risk (in contrast with conventional and biomass generators, which usually attempt to mitigate this risk by indexing the power price by reference to fuel prices);

- a solar Generator suffers from intermittency risk subject, usually but not necessarily, to a combination of the following mitigants: (i) a "take-and-pay" obligation applies to the whole plant's output (rather than just a percentage of it); (ii) if the plant's output is "curtailed" for any reason not attributable to the

Generator, power would still be paid for on a "deemed" delivery basis; (iii) no "minimum dispatch" obligation applies; (iv) possible imbalance risk (ie the risk that the grid operator may charge penalties for (usually irradiation related) deviations between the power nominated and that actually injected into the grid) is borne by the Offtaker; (v) in some markets (eg EU) "green" power benefits from a "priority of dispatch" thus reducing the risk of curtailment by the relevant transmission system operator (TSO); and

- contract consideration (ie the energy charge) is usually variable (ie related to the actual kWh produced) and no fixed capacity (or availability) charge applies.

The main reasons underpinning the solar Generator's need to enter into a PPA may include the following:

- ensuring a subsidised tariff for power produced when grid parity has not yet been achieved;
- ensuring a route to market (given the intermittency of the solar source and/or in markets where no priority of dispatch for renewables sources applies);
- a monopolistic state-owned Offtaker purchases all electricity produced (this is often the case in emerging markets);
- the Offtaker wishes to ensure security of supply (this is often the case in markets with congested or fragmented grids or with insufficient national production) and, more importantly
- Lenders require a predictable cash flow in order to finance the construction of the solar plant on a limited recourse basis.

For obvious reasons, a PPA alone is not sufficient to secure a predictable cash flow where an Offtaker has a poor credit rating.

Therefore, Lenders will pay a great deal of attention to the analysis of the Offtaker's creditworthiness. To this end, credit enhancement and performance securities (eg parent company or sovereign guarantees, letters of credit, cash collaterals or prepayments) may be required if the Offtaker's financial robustness is questioned, particularly in emerging markets where an alternative route to market (eg an electricity pool) is not available in the event of Generator's default.

Typical Main Terms

Construction, testing and connection related issues. Given the Offtaker's interest in timely purchasing the power produced by the solar plant (at times also due to the possible expiry of available incentive schemes), a PPA should provide for (capped) delay liquidated damages (LDs) accruing on the Generator if the planned take over of the plant is delayed, provided these LDs may not exceed those payable under the Engineering, Procurement and Construction Contract (EPC Contract), taking also into account debt service requirements. Force majeure provisions would normally excuse a Generator's delay, provided that if take over of the plant does not occur by an agreed longstop date, the Offtaker may opt to terminate the PPA.

At times, in emerging markets, Generators have also been able to successfully negotiate the inclusion of permitting risk as a force majeure event. Before achieving commercial operations the Generator will need to satisfy the Offtaker that the solar plant has successfully passed certain performance tests.

In this respect, the same set of tests should ideally apply both under the PPA and the EPC Contract. Failure to achieve the guaranteed performances upon testing would normally result in performance LDs being payable to the Offtaker (again up to a capped amount not exceeding LDs payable under the EPC Contract), provided that there is no reduction in the energy charge payable to the Generator as this would result in a double hit.

Finally, the PPA must clearly set out which party bears the costs and risk of connecting the solar plant to the grid. Clearly, the more significant these risks (particularly in emerging markets), the more the Lenders will require the Offtaker to bear them. If the Offtaker is responsible for the connection and it does not timely complete the relevant facilities, the solar plant will nonetheless be deemed available and this will result in the Offtaker's obligation to commence payment of the energy charge.

Conditions precedent and duration of agreement. Generally, a PPA takes effect after all applicable conditions precedents (CPs) have been met (or waived). Most commonly used CPs include obtainment of all permits, execution of all major project contracts, taking out of all required insurances, and, more importantly, achievement of financial close. PPAs are usually long-term contracts (15 to 25 years from the commercial operations date) because duration must be sufficient to repay any required lending and produce required Sponsors' return. In particular, the Sponsors will want to see a PPA that is consistent with the design life of the solar plant or, at least, the incentives period. Duration is usually extended, on a day-by-day basis, by periods of Generator's force majeure.

Quantities and dispatch risk. The Offtaker must purchase, take delivery of and pay for 100% of the solar plant's output. Enforceability of "take-or-pay" and "take-and-pay" provisions is a sensitive issue, which must be investigated with lawyers. If the electricity cannot be physically offtaken as it is "curtailed" due to grid constraints, emergency conditions, voluntary actions by TSO or force majeure events occurring outside the solar plant's battery limits (possibly including permitting issues not attributable to the Generator), electricity will still be calculated and paid for on a "deemed" delivery basis.

As illustrated above, given the intermittency of the solar source, Lenders will most likely require that the Offtaker also bears any imbalance risk.

However, in light of the recent evolution of weather forecasting models, there are instances (particularly in developed energy markets) in which Generators (and their Lenders) may be available to accept such risk, provided that the relevant liabilities are capped.

Price and foreign exchange risk. A solar PPA normally provides for a unit price per kWh (variable charge). Therefore, price is related to the actual electricity delivered (or deemed delivered in the event of curtailment).

The variable charge may be flat or escalate over time but, in any event, it must be sufficient to pay both for Generator's fixed costs (ie debt service and fixed operating and maintenance costs) and Sponsors' equity return. In this vein, any local law provision contemplating the possibility to re-open the price (eg hardship clauses) should be specifically contracted out of.

A PPA must also clearly define each party's ownership to ancillary products – and entitlement to the related revenues – arising out under local incentivising schemes (eg renewables related certificates, certificates of origin) or grid code rules (eg capacity payment, grid services).

Finally, a Generator should be exempted from any foreign exchange risk. Therefore, if commercial hedging is not available, an Offtaker will normally be expected to take this risk and the variable charge should be denominated in (or linked to an exchange rate of) the currency of the Generator's debt. Also, there should be no restriction to the transferability of the proceeds from the host country to offshore accounts.

Measurement and payment terms. Typically, the Generator is responsible for installing, maintaining, periodically check, repair and replace (if necessary) all metering equipment in accordance with best industry practice (and applicable grid code rules) so as to determine the delivered power and enable periodical issue of detailed invoices. The Offtaker will then be bound to pay the invoiced amount without any set-off (other than in respect of LDs possibly accruing on the Generator) or deduction (other than in case of manifest error or, at times, *bona fide* dispute). Invoices should be deemed final if not disputed within an agreed period of time.

Force majeure. The Generator will not be deemed available (and, therefore, no energy charge will be payable) during periods of force majeure affecting the Generator to the extent that the relevant event would prevent or restrict the solar plant's operation.
At times, the Generator may be entitled to consequently extend the duration of the PPA on a day per day basis.
Moreover, the Generator may, to some extent, be able to take insurance against the adverse effects stemming from certain force majeure events. On the other hand, the Generator will be deemed available (and, therefore, the electricity would be deemed delivered and paid for) during periods of other types of force majeure (notably force majeure affecting the grid or the Offtaker and, at times, permitting or change in law issues).
Either party is usually entitled to terminate the PPA for prolonged force majeure (typically lasting more than 18-24 months) which renders performance of the PPA impossible.

Change in circumstances. A change in tax is usually considered an Offtaker's risk, thus resulting in a corresponding increase of the energy charge, only to the extent that it is intended to discriminate the Generator or its Sponsors.
Therefore, a change in tax applying across the industry (eg an increase of the corporate income tax) is usually considered as part of the Generator's normal entrepreneurial risk. On the other hand, the Generator should be kept whole through an increase of the energy charge if a change in law adversely affects its profits. Moreover, in instances where a change in law requires a capital expenditure by the Generator (eg laws increasing the solar plants decommissioning and modules disposal standards) and the latter, acting in good faith, is not able to raise the necessary debt, the Offtaker may be required to act as a "lender of last resort", particularly in emerging markets.

A softer approach may apply to a mere change in industry rules not resulting in lower Generator's profits. In this event, the parties should merely negotiate in good faith the amendments required to maintain the initial commercial underpinning of the PPA.

Default and termination. A PPA should clearly state the events of default on the basis of which a party may, subject to a cure period, terminate the contract (eg delay in achieving commercial operations, poor availability or performance over a certain period of time, failure to pay amounts when falling due, failure to provide an agreed security, insolvency) and the relevant consequences.

These vary depending on the triggering event. First, if a PPA is terminated due to Offtaker's default, the Generator must receive the net present value (NPV) of all energy charges which would have been payable if the PPA had not been terminated, so as to enable the Generator to repay the outstanding loan and compensate the Sponsors for their losses.

This is subject, at times, to a mitigation clause under which the profits that the Generator could make through merchant operation of the solar plant (quite obviously in markets where this is possible) are deducted from the NPV. Second, if a PPA is terminated due to prolonged force majeure, then the Generator should receive a termination amount equal to the outstanding debt.

Finally, in the event of termination due to Generator's default, no compensation should apply given that the Lenders have the possibility to save the project by enforcing their securities and/or exercising their step-in rights. The possibility to terminate due to insolvency events and the applicable consequences are, again, an area where local law advise is strongly recommended.

Limitation of liability. For bankability reasons, a PPA normally excludes that either party could be liable for consequential or indirect losses save in the case of wilful misconduct or gross negligence.

Assignment and Direct Agreement. Assignment by either party is not permitted save that the Generator may create security interests required by the Lenders in respect of the Generator's proceeds.

Lenders normally require a direct contractual relationship with the Offtaker which:

      a- requires the Offtaker to notify them before terminating the PPA and to afford them a reasonable opportunity of remedying any breach (cure period);

b- prevents the Offtaker from terminating the PPA solely by reason of an enforcement by the Lenders of their security; and

c- makes it clear that, in either (a) or (b) above, the Lenders may take over the running of the solar plant or transfer it to a third party (step-in right).

Governing law and dispute resolution. Differently from finance documents, it is commonly accepted that domestic law applies to a PPA (as well as to other project contracts). This is subject both to a local lawyer's "sanity check" ensuring that "what you see is what you get" and, in the event of state-owned Off takers, also to a waiver of sovereign immunity. On the other hand, particularly in emerging markets, a dispute resolution clause providing for an arbitration to be conducted in a neutral seat under generally accepted rules (eg ICC or UNICTRAL) would normally be considered of paramount importance by both Sponsors and Lenders.

18.4) *Valuation of economics*

<u>Background</u>

Capital budgeting and power generation management become the most critical and success factors of solar business plan to be performed.

In this way, industrial cash flows and project optimization as well as commercial and technical managements are strictly linked each other.

In the light of value addition for the company and its shareholders, project developer as well as top management should remember to maximize the difference between invested capital performance and its weighted average cost, especially as corporate benefit:

$$EVA \% = ROI \text{ after tax} - WACC$$

Where, according International glossary, they are to be definied as:
- ✓ EVA: Economical Value Added
- ✓ ROI: Return on Investment
- ✓ WACC: Weighted Average Cost of Capital

In front of such a basic corporate principle, the business concept to the project investment and execution results shall be the "Value Based Management" or V.B.M. Thus, all the efforts are focused to increase the following range, especially as shareholder benefit:

$$TSR - ke$$

where they mean:
- ▪ TSR: Total Shareholder Return
- ▪ Ke: Cost of Equity

In order to make a proper measurement of VBM, it is crucial that a valuation tool could calculate and compare TSR vs. EVA, which must be carried out.

Operating Managers shall inspect, monitor and control each specific business stimulus and financial tool to orient all variables for the value creation and addition, concerning operationally:

⇒ EBIT through more efficiency, cost reduction, process optimization.

⇒ Invested Capital through proper financial planning, reduction of capital needs.

⇒ Kd through debt cost reduction, optimization of financial *leverage*.

⇒ Ke through Country and Market risks assessment, as well as business and technology risks.

Such scheme could illustrate that several parameters and additional functions are to be considered and fixed, under a complete valuation process:

$$
\text{EVA}
\begin{cases}
\text{ROI}
\begin{cases}
\text{EBIT} \\
\text{Invested Capital}
\end{cases} \\
\text{WACC}
\begin{cases}
\text{Kd} \\
\text{Ke}
\end{cases}
\end{cases}
$$

Where:

- ROI come from [EBIT/Inv. Capital]
- WACC = Ke (E + D)/E + Kd (D + E)/D

from which:

✓ D/E or Debt/Equity ratio shows the financial leverage of project SPV.

✓ Ke is the term of equity cost (project financing/bank risk side)

✓ Kd means the debt cost (shareholder risk side).

18.4) Valuation Model

One of the most important question in corporate finance modeling is understanding the value toady of a cash flow to be received at a later date.

Among different mathematical approach and modeling, the best way to finally valuate a solar project and/or an operating solar plant asset is exactly DCF or discounted cash flows methodology.

First of all, the basic step of any mathematical and financial valuations shall start up from a complete due diligence phase of the solar project characteristics.

From industrial point of view, there are four focal points/documents to be checked and assessed under the best practice of valuation for the final decision makers:

1. Assumption Book
2. Results and Sensitivity Analysis
3. Profit&Loss and Cash Flows: unlevered basic case
4. Profit&Loss and Cash Flows: levered simulation

Therefore, key economics and financial indicators must be calculated and synthesized in the following parameters of business and investment performance:

➤ NPV or Net Present Value
➤ IRR or Internal Rate of Return
➤ PBP or Pay Back Period
➤ PI or Profitability Index

DCF method works through considering disbursement of initial capital $[-CF_0]$ as well as the cash flows $[+CF_n]$ related to the incomes of [n] subsequent years:

$$+ CF_1 + CF_2 + CF_3 + CF_4 ..... + CF_n$$

In that way, the Net Present Value can be easily rendered as the following mathematical function:

$$NPV(WACC, n) = - CF_0 + CF_1/(1 + WACC)^1 + CF_2/(1 + WACC)^2 + ... + CF_n/(1 + WACC)^n$$

A positive NPV amount is directly an expression of the measure of value created; in opposite, a negative NPV is going to quantify the value destroyed, coming from the effect of capital investment.

IRR is the key indicator to index financial profitability and business performance of project, as the compound annual growth rate of investment return.

Its specific calculation, as discount rate, needs to run on iterative cycles in order to output NPV = 0 of cash flows.

Thus, IRR value shall depend on both: amount as well as temporal distribution curve of cash flows, generated by capital investment (Capex).

From mathematical side, it showed from the following polynomial equation:

$$NPV = - CF_0 + CF_1 /(1 + IRR)^1 + CF_2 /(1 + IRR)^2 + \ldots.+ CF_n/(1 + IRR)^n = 0$$

According the best practice of valuation, also PBP should be used to demonstrate to the investment committee how many years project needed, so that cumulated cash flows shall reach the initial Capex.

That means to solve the following calculation, in order to complete with the other indicators the valuation picture of economics, based on every specific project business plan:

$$PBP = CF_1 + CF_2 \ldots.+ CF_{n-t} - CF_0 = 0$$

BIBLIOGRAPHIC REFERENCES

1. O. Bramanti et Al., *"Le Tecnologie delle Fonti Energetiche Rinnovabili"*, Sole24Ore Technology-Professional Edition, December 2012.

2. W. Ratismith, A. Inthongkhum, *"A Novel Non-Tracking Solar Collector for High Temperature Application"*, Proceedings of the 25th international conference on efficiency, cost, optimization, simulation and environmental impact of energy systems, Perugia, 26-29 June 2012.

3. D. Cano, J.M. Monget, M. Albuisson, H. Guillard, N. Regas, L. Wald (1986), *"A Method for the Determination of the Global Solar Radiation from Meteorological Satellite Data"*, Solar Energy, 37, pp. 31–39, Elsevier.

4. Erbs D.G., Klein S.A. and Duffie J.A. (1982), *"Estimation of the Diffuse Radiation Fraction for Hourly, Daily and Monthly Average Global Radiation"*, Solar Energy, vol. 28, 1982, pp. 293-302.

5. M. Iqbal (1983), *"An Introduction to Solar Radiation"*, Academic Press Canada, Don Mills (Ontario, Canada), 1983, ISBN: 0-12-373750-8, pp. 390.

6. PV News, 2012-2014.

7. Il mercato mondiale del PV a concentrazione (CPV), IMS Research Report, settembre 2012.

8. Brett Prior, Chitra Seshan, Concentrating Photovoltaics 2011: Technology, Costs and Markets, GMT Research Report, May 2011.

9. Green M.A., Silicon Solar Cells. Advanced Principles and Practice, University of South Wales, Sydney, 1995

10. Sze S.M., Semiconductor Devices, John Wiley &Sons, Hoboken, 1985

11. Luque A. & Hagedus S. eds., Handbook of Photovoltaic Science & Engineering, John Wiley & Sons, Hoboken, 2003.

12. Martin A. Green et al., Cell Efficiency Tables (version 40), Prog. in Photovoltaics: Res. Appl., 2012, 20.

13. Chopra K. Et al., Thin Film Solar Cells: an Overview, Progress in Photovoltaics: Res. Appl., 2004/1.

14. *SolarItaly "Italian Atlas of Solar Radiation"*, Internet website address: http://www.solaritaly.enea.it.

15. *Clisun "Climate Archive ENEA-DBT"*, Internet website address: http://clisun.casaccia.enea.it.

16. E. Cogliani, D. Malosti, M. Mancini e S. Petrarca (1993), *"Estimating Global Solar Radiation on the Soil Using Secondary Images by Meteosat"*, HTE Alternative energies, 85, pp. 268-273.

17. S. Petrarca, E. Cogliani, F. Spinelli (2000), *"Global Solar Radiation on the Soil in Italy"*. Years 1998 and 1999 and 1994-1999 average, ENEA, Rome.

18. F. Spinelli, E. Cogliani, A. Maccari, M. Milone (2007), *"Measuring and Estimating Solar Radiation: ENEA Archives and the Internet site of the Italian Atlas of Solar Radiation for data publication"*, Technical report ENEA SOL/RS/2007/21, Rome.

19. Boland J. and Ridley B. (2008), *"Models of Diffuse Solar Fraction, in Modeling Solar Radiation at the Earth's Surface"*, Springer-Verlag Berlin Heidelberg, cap. 8, pp. 193-219.

20. European Organisation for the Exploitation of Meteorological Satellites (EUMETSAT), web site: http://www.eumetsat.int.

21. R.R. Perez, R. Stewart, C. Arbogast, R. Seals, J. Scott (1986), *"An Anisotropic Hourly Diffuse Radiation Model for Sloping Surfaces – Description, Performance Validation, Site Dependency Evaluation"*, Solar Energy 36, 6 (1986), pp. 481-497.

22. R.R. Perez, P. Ineichen, R. Seals, J. Michalsky, R. Stewart (1990), *"Modeling Daylight Availability and Irradiance Components from Direct and Global Irradiance"*, Solar Energy 44, 5 (1990), pp. 271-289.

23. M. A. Cucumo, V. Marinelli, G. Oliveti (1994), *"Solar Engineering.* Principles and Applications"*, Pitagora Editrice Bologna, ISBN: 88-371-0729-3, pp. 422.

24. Y. Cascone, V. Corrado, V. Serra, C. Toma (2010), *"Calculation of Shadowing the Building Cover"*, Technical report, prepared for the event "Electrical System Research" Ministry for Economic Development, Rome.

25. F. Lanini (2010), *"Division of Global Radiation into Direct Radiation and Diffuse Radiation"*, Master's thesis, Faculty of Science, University of Bern.

26. Francesco Spinelli, Euro Giovanni Cogliani, Augusto Maccari, Mauro Milone, *"La misura e la stima della radiazione solare: l'archivio dell'ENEA e il sito Internet dell'Atlante Italia o della radiazione solare per la pubblicazione dei dati"*, Rapporto Tecnico ENEA/SOL/RS/2007/21.

27. Massimo Falchetta et al., *"Il programma ENEA sull'energia solare a concentrazione ad alta temperatura"*, Rapporto Tecnico ENEA/SOL/RS/2005/22.

28. Fabrizio Fabrizi et al., *"Impianto P.C.S. - Componenti e apparecchiature di processo – Progetto esecutivo-"*, Rapporto Tecnico ENEA/SOL/RS/2002/25.

29. Pietro Tarquini et al., *"Fluido termovettore: dati di base della miscela di Nitrati di Sodio e Potassio"*, Rapporto Tecnico ENEA/SOL/RS/2001/07.

30. Antonio De Luca e al, *"Modello di scambio termico allo stato stazionario e transitorio del tubo ricevitore e analisi dei dati sperimentali sui tubi SHOTT"*, Rapporto Tecnico ENEA/SOL/RS/2005/20.

31. Fabrizio Fabrizi et al, "*Analisi dati sperimentali campagna di prove 5 aprile 2004 – 19 settembre 2004*", Rapporto Tecnico ENEA/SOL/RS/2004/27.

32. W. Kostowski, J. Skorek "*Thermodynamic and economic analysis of heat storage application in co-generation systems*", Int. J. Energy Res. 2005, 29, 177-188.

33. Mauro Vignolini, Giorgio Simbolotti, Alfredo Fontanella, "*Il progetto solare termodinamico dell'enea: un importante risultato scientifico e una opportunità per l'economia italiana*", Energia, Ambiente e Innovazione, 2/2010.

34. Mauro Vignolini, Alfredo Fontanella, "*Energia dal deserto – I grandi progetti per le rinnovabili nel Mediterraneo*", cap. 21 "Solare termodinamico", 2011, Edizioni Ambiente

35. Tommaso Crescenzi, Alfredo Fontanella, "*Italian research on Concentrated Solar Power, 2011*", Paper for the Web Conference ENEA in Japan.

36. Fabiola Falconieri, Massimo Maffioletti, Alfredo Fontanella, "*Test facility for the development of CSP*", 2011, Video Podcast for the Web Conference ENEA in Japan.

37. Francesco Di Mario, Alfredo Fontanella, "*Quaderno sul Solare Termodinamico*", ENEA, 2011 (http://www.enea.it/it/enea_informa).

38. Mauro Vignolini, Alfredo Fontanella, "*ENEA Thermodynamic Solar Project*", 2010, Forum Italia – Cina per l'innovazione.

39. Mauro Vignolini, Alfredo Fontanella, "*Energia solare a concentrazione – Piano strategico 2010*", Rapporto tecnico ENEA.

40. Mauro Vignolini et al., "*Progetto Archimede: Realizzazione di un impianto solare termodinamico integrativo presso la Centrale ENEL di Priolo Gargallo (SR)*", Rapporto Tecnico ENEA/SOL/RS/2007/14.

41. AA.VV. "*Concentrating Solar Power (CSP)*", NREL Report (www.nrel.gov/csp/publications.html).

42. AA.VV., "*Global Concentrating Solar Power – Outlook 2009*", SolarPaces.

43. Francesco Di Mario, Alfredo Fontanella, "*La ricerca dell'ENEA sul solare a concentrazione (CSP)*", INGENIO, agosto 2012.

44. Massimo Falchetta, Augusto Maccari e Mauro Vignolini, "*Il sole concentrato*", Le Scienze, n. 459, novembre 2006.

45. Global Market Outlook for Photovoltaics until 2016, EPIA Report, May 2012

46. www.solarbuzz.com, March 2012.

47. Madan A. & Shaw M.P., The Physics and Application of Amorphous Semiconductors, Academic Press, 1988.

48. Steabler D.L. & Wronsky C.R., J. Appl. Phys. 51, 3262, 1980.

49. Schropp R.E.I & Zeman M., Amorphous and Microcrystalline Silicon Solar Cells: Modelling, Materials and Device Technology, Kluwer, Boston, 1988.

50. Yang J. et al., Triple-Junction Amorphous Silicon Alloy Solar Cells with 14.6% Initial and 13.0% Stable Conversion Efficiencies, Applied Physics Letters, 2975-2977, 1997/70.

51. Shah A. et al., Microcrystalline Silicon and "Micromorph" Tandem Solar Cells, Thin Solid Films, 403-404, 2002/179.

52. Vossen J.L. & Kern W., Thin Film Processes, Academic Press, 1978.

53. Dhere, Neelkanth G., Toward GW/year of CIGS Production within the next Decade, Solar Energy Materials and Solar Cells 91 (15–16): 1376. 2007.

54. Stanbery B.J., Critical Reviews in Solid State and Materials Science 27: 73, 2002.

55. Kurtz S., Opportunities and Challenges for Development of a Mature Concentrating Photovoltaic Power Industry, Technical Report, NREL/TP-5200-43208-2012.

56. Reisfeld R., New developments in luminescence for solar energy utilization, Optical Materials 32 (9): 850–856, 2010.

57. Branker K., Pathak M.J.M., Pearce J.M., Energy Reviews, Volume 15, Issue 9, December 2011, 4470-4482.

58. http://www.energy.eu/#Domestic-Elec.

59. Califano F. et al., La progettazione dei sistemi fotovoltaici, Liguori, Napoli, 1988.

60. Moroni M., Nitrati G., Progettazione fotovoltaica in conto energia, EPC libri, Roma, 2007.

61. Green M.A., Third Generation Photovoltaics: Ultra High Conversion Efficiency at Low Cost, in Progress in Photovoltaic: Research and Applications vol. 9, p. 123, 2001.

62. www.sj-solar.com.

63. Hanna M.C., Nozik A.J.J., Appl. Phys. 2006, 100.

64. Klimov V.I., Appl. Phys. Lett. 2006, 89.

65. Sam-Shajing Sun, Niyazi Serdar Sariciftci, Organic Photovoltaics - Mechanisms, Materials, and Devices, Taylor & Francis Group, 2005.

66. Graetzel O'Regan, A Low-cost, High-efficiency Solar Cell Based on Dye-sensitized Colloidal TiO_2 Film, Nature, 1991, Vol. 353.

67. Dennler G. et al., Polymer-Fullerene Bulk-Heterojunction Solar Cells, Adv. Mater., 2009, 21, 1323–1338.

68. Todorov T.K. et al., Beyond 11% Efficiency: Characteristics of State-of-the-Art $Cu_2ZnSn(S,Se)_4$ Solar Cells. - Advanced Energy Materials, 16 aug. 2012.

69. Xiaochang Miao et al., High Efficiency Graphene Solar Cells by Chemical Doping, Nano Letters, 2012.

70. Becquerel E., Mémoire sur les effets électriques produits sous l'influence des rayons solaires, Comptes Rendus 9: 561–567, 1839.

71. V. Quaschning, *Technology Fundamentals - Solar Thermal Power Plants*, Renewable Energy World Vol. 6 (2003) pp. 109-113.

72. S.M. Jeter, *Maximum conversion efficiency for the utilization of direct solar radiation*, Sol. Energy 26 (1981) 231-236.

73. R. Petela, *Energy of heat radiation*, J. heat Transfer 86, (1964) 187-192. W.H. Press, *Theoretical Maximum for energy for direct and diffuse sunlight*, Nature 264, (1976) 734-735.

74. "Molten Salts systems other applications link to Solar Power Plants". National Renewable Energy Laboratory (NREL). Retrieved 2011-09-06.

75. Types of solar thermal CSP plants. Tomkonrad.wordpress.com. Retrieved on 22 April 2013.

76. Molten salt as CSP plant working fluid. (PDF). Retrieved on 22 April 2013.

77. Control of Solar Energy Systems, Eduardo F. Camacho, Manuel Berenguel, AIC SPRINGER.

78. Design and simulation of the solar chimney power plants with TRNSYS. Fei Cao, Huashan Li, Liang Zhao Tianyang Bao, Liejin Guo. Solar Energy 2013.

79. Schlaich J, Schiel W (2001), "Solar Chimneys", in RA Meyers (ed), *Encyclopedia of Physical Science and Technology, 3rd Edition*, Academic Press, London.

80. Solar Thermal Power Plants – On the Way to Commercial Market Introduction, Prof. Hans Muller-Steinhagen, Institute for Technical Thermodynamics, German Aerospace Centre (DLR), Stuttgart-Cologne-Almeria/Spain.

81. Chang, Yi-Lu; Lu, Zheng-Hong (2013). "White Organic Light-Emitting Diodes for Solid-State Lighting". *Journal of Display Technology* PP (99).

82. "Bringing the DESERTEC vision into reality". Dii GmbH. Retrieved 24 Dec. 2010.

83. R. L. Arántegui, N. Fitzgerald and P. Leahy, "Pumped-Hydro Energy Storage: Potential for Transformation from Single Dams," JRC Institute for Energy and Transport, EUR 25239 EN, 2011.

84. Modeling Solar Radiation at the Earths Surface, Viorel Badescu, Springer.

85. Andrade da Costa, B., Lemos, J.M.: An adaptive temperature control law for a solar furnace. Control Eng. Pract. 17, 1157–1173 (2009).

86. Almorox, J., Hontoria, C.: Global solar radiation estimation using sunshine duration in Spain. Energy Convers. Manag. 45, 1529–1535 (2004).

87. Renewables 2013, Global Status Report 2013, REN21, Renewable Energy Policy Network for the 21st Century.

88. Duffie J. A., Beckman W.A. (2006), Solar Engineering of Thermal Processes, 3rd edition, John Wiley and Sons.

89. Goswami D.Y., Kreith F., Kreider J.F. (1999), Principles of Solar Engineering, 2nd edition, Taylor & Francis.

90. Cucumo M.A., Marinelli V., Oliveti G. (1994), Ingegneria solare principi e applicazioni, Pitagora Editrice.

91. ENEA (2009), Dossier "Usi termici delle fonti rinnovabili", Workshop "Usi termici delle fonti rinnovabili", 11 novembre 2009, Roma, ENEA Ed. , 75 pp.
92. ENEA (2010), Dossier "Fonti rinnovabili. Competenze e infrastrutture ENEA", luglio 2010.
93. ENEA (2011), Quaderno "Solare termico a bassa e media temperatura", luglio 2011.
94. Lazzarin R. (2011), Pompe di calore, SGEditoriali.
95. Setterwall F. (2002), Technical grade paraffin waxes as phase change materials for cool thermal storage system capital cost estimation, Energy Conversion and Management 43 (13), pp. 1709-1723, www.sciencedirect.com.
96. Lazzarin R., Busato F., Noro M. (2011), Studio di sistemi di accumulo a cambiamento di fase nel solar cooling.
97. Sharma A., Tyagi V.V., Chen C.R., Buddhi D. (2007), Review on thermal energy storage with phase change materials and applications, Renewable and Sustainable Energy Reviews 13 (2), pp. 318-345, www.sciencedirect.com.
98. Villarini M., Germanò D., Fontana F., Limiti M. (2010), Sistemi solari termici per la climatizzazione, Maggioli Editore.
99. Calabrese N., Trinchieri R., Simonetti A. (2011), Utilizzo dell'energia elettrica e solare per condizionamento estivo.
100. Energy & Strategy Group (2012), Solar Energy Report, Politecnico di Milano, www.energystrategy.it.
101. Pauschinger T. (2002), Impianti solari termici - Manuale per la progettazione e costruzione, Ambiente Italia.

www.ingramcontent.com/pod-product-compliance
Lightning Source LLC
Chambersburg PA
CBHW050037220326

41599CB00040B/7189